**POPULAR BOOKS**

**ON**

**Natural Science**

# POPULAR BOOKS
# ON
# Natural Science.

FOR PRACTICAL USE IN EVERY
HOUSEHOLD,

FOR READERS OF ALL CLASSES.

By A.. BERNSTEIN

GYAN

*Published by*

Gyan Publishing House
5, Ansari Road
Daryaganj, New Delhi-110002
Phone: 011-47034999, 9811692060
E-mail: books@gyanbooks.com

*Distribution Network*
gyanbooks.com
India, USA, Canada, UK, Australia



ISBN: 978-93-6280-396-2 (HB)
First Published, 1869

2nd Impression 2024

*Printed at:* Gyan Press, Delhi.



**POPULAR BOOKS ON NATURAL SCIENCE**
Author: **A.** BERNSTEIN

# POPULAR BOOKS

ON

# Natural Science.

## FOR PRACTICAL USE IN EVERY HOUSEHOLD,

FOR READERS OF ALL CLASSES.

By A. BERNSTEIN.

---

*CONTENTS:*



---

New York:
CHR. SCHMIDT, PUBLISHER, 39 CENTRE STREET.

# BERNSTEIN'S

# POPULAR TREATISE

ON

# NATURAL SCIENCE.

"In primis, hominis est propria VERI inquisitio atque investigatio. "Itaque cum sumus negotiis necessariis, curisque vacui, tum avemus "aliquid videre, audire, ac dicere, cognitionemque rerum, aut occul- "tarum aut admirabilium, ad benè beatéque vivendum necessariam "ducimus;—ex quo intelligitur, quod VERUM, simplex, sincerumqe sit, "id esse naturæ hominis aptissimum. Huic veri videndi cupiditati "adjuncta est appetitio quædam principatûs, ut nemini parere ani- "mus benè a naturâ informatus velit, nisi præcipienti, aut docenti, "aut utilitatis causâ justè et legitimè imperanti: ex quo animi magni- "tudo existit, et humanarum rerum contemtio."

CICERO, DE OFFICIIS, Lib. 1. § 13.

Before all other things, man is distinguished by his pursuit and investigation of TRUTH. And hence, when free from needful business and cares, we delight to see, to hear, and to communicate, and consider a knowledge of many admirable and abstruse things necessary to the good conduct and happiness of our lives: whence it is clear that whatsoever is TRUE, simple, and direct, the same is most congenial to our nature as men. Closely allied with this earnest longing to see and know the truth, is a kind of dignified and princely sentiment which forbids a mind, naturally well constituted, to submit its faculties to any but those who announce it in precept or in doctrine, or to yield obedience to any orders but such as are at once just, lawful, and founded on utility. From this source spring greatness of mind and contempt of worldly advantages and troubles.

# CONTENTS.

## PART I.

### THE WEIGHT OF THE EARTH.



## PART II.

### VELOCITY.



## PART III.

### NUTRITION.

## PART IV.

### LIGHT AND DISTANCE.



## PART V.

### THE WONDERS OF ASTRONOMY.



## PART VI.

### METEOROLOGY.

## PART VII.

### OUR ARTICLES OF FOOD.

# PART I.

# THE WEIGHT OF THE EARTH.

# CHAPTER I.

## HOW MANY POUNDS THE WHOLE EARTH WEIGHS.

NATURAL philosophers have considered and investigated subjects that often appear to the unscientific man beyond the reach of human intelligence. Among these subjects may be reckoned the question, "How many pounds does the whole earth weigh?"

One would, indeed, believe that this is easy to answer. A person might assign almost any weight, and be perfectly certain that nobody would run after a scale, in order to examine, whether or not an ounce were wanting. Yet this question is by no means a joke, and the answer to it is by no means a guess; on the contrary, both are real scientific results. The question in itself is as important a one, as the answer, which we are able to give, is a correct one.

Knowing the size of our globe, one would think that there was no difficulty in determining its weight. To do this, it would be necessary only to make a little ball of earth that can be accurately weighed; then we could easily calculate how many times the earth is larger than this little ball; and by so doing, we might tell, at one's finger-ends, that—if we suppose the little earth-ball to weigh a hundred-weight—the whole globe, being so many times larger, must weigh so many hundred-weights.

Such a proceeding, however, would be very likely to mislead us. For all depends on the substance the little ball is made of. If made of loose earth, it will weigh little; if stones are taken with it, it will weigh more; while, if metals were put in, it would, according to the kind of metal you take, weigh still more.

If, then, we wish to determine the weight of our globe by the weight of that little ball, it is first necessary to know of what our globe consists; whether it contains stones, metals, or things entirely unknown; whether empty cavities, or whether, indeed, the whole earth is nothing but a hollow sphere, on the surface of which we live, and in whose inside there is possibly another world that might be reached by boring through the thick shell.

With the exercise of a little thought, it will readily be seen that the question, "How much does our earth weigh?" in reality directs us to the investigation of the character of the earth's contents; this, however, is a question of a scientific nature.

The problem was solved not very long ago. The result obtained was, that the earth weighs 6,069,094,272 billions of tons; that, as a general thing, it consists of a mass a little less heavy than iron; that towards the surface it contains lighter materials; that towards the centre they increase in density; and that, finally, the earth, though containing many cavities near the surface, is itself not a hollow globe.

The way and manner in which they were able to investigate this scientifically, we will attempt now to set forth as plainly and briefly as it can possibly be done.

## CHAPTER II.

### THE ATTEMPT TO WEIGH THE EARTH.

It is our task to explain, by what means men have succeeded in weighing the earth, and thus become acquainted with the weight of its ingredients.

The means is simpler than might be thought at the moment. The execution, however, is more difficult than one would at first suppose.

Ever since the great discovery of the immortal Newton, it has been known that all celestial bodies attract one another, and that this attraction is the greater, the greater the attracting body is. Not only such celestial bodies as the sun, the earth, the moon, the planets,and the fixed stars, but *all* bodies have this power of attraction; and it increases in direct proportion to the increase of the mass of the body. In order to make this clear, let us illustrate it by an example. A pound of iron attracts a small body near by; two pounds of iron attract it precisely twice as much; in other words, the greater the weight of an object, the greater the power of attraction it exercises on the objects near by. Hence, if we know the attractive power of a body, we also know its weight. Nay, we would be able to do without scales of any kind in the world, if we were only able to measure accurately the attractive power of every object. This, however, is not possible; for the earth is so large a mass, and has consequently so great an attractive power, that it draws down to itself all objects which we may wish other bodies to attract. If, therefore, we wish to place a small ball in the neighborhood of ever so large an iron-ball, for the purpose of having the little one attracted by the large one, this little ball will, as soon as we let it go, fall to the earth, because the attractive power of the earth is many, very many times greater than that of the largest iron-ball ; so much greater is it, that the attraction of the iron-ball is not even perceptible.

Physical science, however, has taught us to measure the earth's attractive power very accurately, and this by a very simple instrument, viz., a pendulum, such as is used in a clock standing against the wall. If a pendulum in a state of rest—in which it is nearest to the earth—is disturbed, it hastens back to this resting-point with a certain velocity. But because it is started and cannot stop without the application of force, it recedes from the earth on the other side. The earth's attraction in the meanwhile draws it back, making it go the same way over again. Thus it moves to and fro with a velocity which would increase, if the earth's mass were to increase ; and decrease, if the earth's mass were to decrease. Since the velocity of a pendulum may be measured very accurately by counting the number of vibrations it makes in a day, we are able also to calculate accurately the attractive power of the earth.

A few moments' consideration will make it clear to everybody, that the precise weight of the earth can be known so soon as an apparatus is contrived, by means of which a pendulum may be attracted by a certain known mass, and thus be made to move to and fro. Let us suppose this mass to be a ball of a hundred pounds, and placed near a pendulum. Then as many times as this ball weighs less than the earth, so many times more slowly will a pendulum be moved by the ball.

It was in this way that the experiment was made and the desired result obtained. But it was not a very easy undertaking, and we wish, therefore, to give our thinking readers in the next chapter a more minute description of this interesting experiment, with which we shall for the present conclude the subject.

## CHAPTER III.

### DESCRIPTION OF THE EXPERIMENT TO WEIGH THE EARTH.

Cavendish, an English physicist, made the first successful attempt to determine the attractive power of large bodies. His first care was, to render the attraction of the earth an inefficient element in his experiment. He did it in the following way:

On the point of an upright needle he laid horizontally a fine steel bar, which could turn to the right and left like the magnetic needle in a compass-box. Then he fastened a small metallic ball on each end of the steel bar. The balls were of the same weight, for this reason the steel bar was attracted by the earth with the same force at both ends; it therefore remained horizontal like the beam of a balance, when the same weight is lying in each of the scales. By this the attractive force of the earth was not suspended, it is true; but it was balanced by the equality of the weights. Thus the earth's attractive power was rendered ineffective for the disturbance of his apparatus.

Next he placed two large and very heavy metallic balls at the ends of the steel bar, not, however, touching them. The attractive force of the large balls began now to tell; it so attracted the small ones that they were drawn quite near to the large balls. When, then, the observer, by a gentle push, removed the small balls from their resting-place, the large ones were seen to draw them back again. But as the latter could not stop if once started, they crossed their resting-point, and began to vibrate near the large balls in the same manner as a pendulum does, when acted upon by the attractive force of the earth. Of course this force was exceedingly small, compared with that of the earth; and for that reason the vibrations of this pendulum were by far slower than those of a common one. This could not be otherwise; and from the slowness of a vibration, or from the small number of vibrations in a day, Cavendish computed the real weight of the earth.

Such an experiment, however, is always connected with extraordinary difficulties. The least expansion of the bar, or the unequal expansion or contraction of the balls, caused by a change of temperature, would vitiate the result; besides, the experiment must be made in a room surrounded on all sides by masses equal in weight. Moreover, the observer must not be stationed in the immediate neighborhood, lest this might exercise attractive force, and by that a disturbance. Finally, the air around must not be set in motion, lest it might derange the pendulum; and lastly, it is necessary not only to determine the size and weight of the balls, but also to obtain a form spherical to the utmost perfection; and also to take care that the centre of gravity of the balls be at the same time the centre of magnitude.

In order to remove all these difficulties, unusual precautions and extraordinary expenses were necessary. Reich, a naturalist in Freiberg, took infinite pains for the removal of these obstacles. To his observations and computations we owe the result he transmitted to us, viz.: that the mass total of the earth is nearly five and a half times heavier than a ball of water of the same size; or, in scientific language: The mean density of the earth is nearly five and a half times that of water. Thence results the real weight of the earth as being nearly fourteen quintillions of pounds. From this, again, it follows that the matter of the earth grows denser the nearer the centre; consequently it cannot be a hollow sphere.

If we consider, that from the earth's surface to its centre there is a distance of 3,956 miles, and that, with all our excavations, no one has yet penetrated even five miles, we have reason to be proud of investigations which, at least in part, disclose to man the unexplorable depths of the earth.

# PART II.

# VELOCITY.

# CHAPTER I.

## VELOCITIES OF THE FORCES OF NATURE.

In former times, when a man would speak of the rapidity with which light traverses space, most of his hearers thought it to be a scientific exaggeration or a myth. At present, however, when daily opportunity is afforded to admire, for example, the velocity of the electric current in the electro-magnetic telegraph, every one is well convinced of the fact, that there are forces in nature which traverse space with almost inconceivable velocity.

A wire a mile in length, if electrified at one end, becomes in the very instant electrified also at the other end. This and similar things every one may observe for himself; then, even the greatest sceptic among you will clearly see, that the change—or "electric force"—which an electrified wire undergoes at one end, is conveyed the length of a mile in a twinkle, verily as if a mile were but an inch.

But we learn more yet from this observation. The velocity with which the electric force travels is so great, that if a telegraph-wire, extending from New York to St. Louis and back again, is electrified at one end, the electric current will manifest itself at the other end in the same moment. From this it follows, that the electric force travels with such speed as to make a thousand miles in a space of time scarcely perceptible. Or, in other words, it travels a thousand miles in the same imperceptible fraction of a moment that it does a single mile.

And experience has taught us even more yet. However great the distance connected by a telegraphic wire may be, the result has always been, that the time which electricity needs to run that distance, is imperceptibly small ; so that it may well be said, its passage occupies an indivisible moment of time.

One might even be led to believe that this is really no "running through"—in other words, that this transmission of effect from one end of the wire to the other end does not require any time at all, but that it happens, as if by enchantment, in one and the same instant. This, however, is not the case.

Ingenious experiments have been tried, to measure the velocity of the elective force. It is now undoubtedly proved, that it actually does require time for it to be transmitted from one place to another ; that this certain amount of time is imperceptible to us for this reason, viz., that all distances which have ever been connected by telegraph, are yet too small, to make the time it takes for the current to go from one end to the other, perceptible to us.

Indeed, if our earth were surrounded by a wire, it would still be too short for common observation, because the electric force would run even through this space—twenty-five thousand miles very nearly—in the tenth part of a second.

Ingenious experiments have shown that the electric current moves two hundred and fifty thousand miles in a second. But how could this have been ascertained ? And are we certain that the result is trustworthy ?

The measurements have been made with great exactitude. To those who are not afraid of a little thinking, we will try to represent the way in which this measurement was taken ; although a perfect representation of it is very difficult to give in a few words.

# CHAPTER II.

## HOW CAN THE VELOCITY OF THE ELECTRIC CURRENT BE ASCERTAINED.

In order to illustrate, how the velocity of the electric current can actually be measured, we must first introduce the following:

Whenever a wire is to be magnetized by an electric machine, at the moment it touches the machine, a bright spark is seen at the end of the wire. The same spark is seen also at the other end of the wire, if touching another apparatus. Let us call the first spark the "entrance-spark," the other the "exit-spark." If a wire, many miles in extent, is put up, and led back to where the beginning of the wire is, both sparks may be seen by the same observer.

Now it is evident, that the exit-spark appears after the entrance-spark just as much later, as the time it took the electric current to run from one end of the wire to the other end. But in spite of all efforts made, to see whether the exit-spark actually appears later, the human eye has not been able to detect the difference. The cause of this is partly owing to the long duration of the impression upon the retina, which leads us to the belief, that we see objects much longer than we really do; partly, the immense rapidity with which the exit-spark follows the entrance-spark. From these two causes, we are tempted to believe both sparks to appear at the same moment.

By an ingenious and excellent means, however, this defect in our eye has been greatly diminished. It is well worth the trouble to read a description of the experiment attentively. The truly remarkable way in which it was tried, will please all who read it.

In order to measure the velocity of the electric current, the ends of a very long wire are placed one above the other. If, now, one makes the observation with the naked eye, both sparks will be found to stand in a vertical line, one above the other, as the points of a colon, thus (:).

But he who wishes to measure the velocity of the electrical current does not look upon the sparks with the naked eye, but into a small mirror, which, by a clock-work, is made to revolve upon an upright axis with exceedingly great rapidity. Thus he can see both sparks in the mirror. If the apparatus be a good one, it will be observed that the sparks, as seen by the aid of the mirror, do not stand in a vertical line above one another, but obliquely, thus (.·).

Whence does this come?

The reason of it is, that after the appearance of the entrance-spark it takes a short time, before the exit-spark appears. During this short time the mirror moves, though but little, and in it the exit-spark is seen as if it had moved aside from the entrance-spark.

Hence, it is through the movement of the mirror that the time, which is necessary for electricity to go through the circuit of the wire, is ascertained. A little reflection will readily convince the reader, that the time may be precisely calculated, provided three things be known, viz.: the length of the wire, the velocity of rotation of the mirror, and the angular distance of the two sparks as seen in the mirror. Thus: Suppose the wire to be 1,000 miles long; and suppose the mirror is made to revolve 100,000 times in a second. Now, if the electrical current traversed these 1,000 miles of wire during *one* revolution of the mirror, then it follows, that the current must move 1,000 miles in the $\frac{1}{100}$ part of a second; or, 100,000 miles in a second.

It is found, however, that the mirror does not revolve an entire circle, or 360 degrees, while the current is passing over 1,000 miles of wire, but we find that the mirror turns through 144 degrees very nearly; therefore the electric current must travel more than 100,000 miles a second. How much more? Just as many times 100,000 miles, as 144 degrees are contained in 360 degrees (the entire circle), viz., two and a half times. Hence, the current travels 250,000 miles in a second.

# PART III.

# NUTRITION.

# CHAPTER I.

## NOTHING BUT MILK.

CONCEIVE a man, gifted with the keenest intellect, but not knowing from experience, that sucklings grow and become men, and imagine what he would say, if you were to tell him this :

" Know, that the little being you see here, is a suckling, that is, a developing human being, who by and by will become thicker and taller. The bones of his body will become firmer and longer. The muscles that animate these bones will likewise increase in size. The same will happen with regard to his eyes, ears, nose, mouth ; to his head, body, and feet ; every component part of his small body will be developed further and further, until the child will become a perfect man."

There is no doubt, that he who does not know all this from experience, will shake his head at it.

But if you were to tell him : " This development and growth have their source in the baby's sucking at the mother's breast a white juice called milk, and out of this milk all the constituent parts of the child are manufactured within himself,"—certainly your hearer would laugh in your face, and perhaps call you a credulous fool.

" What !" he would exclaim, " do you mean to say that milk contains flesh ? Or can you make bones out of milk, or hair ? Can you make nails and teeth out of milk ? Do you wish to persuade me, that milk may be changed into eyes ? that from milk may be manufactured feet, hands, cheeks, eyelids, and the various other parts of the human body?"

And if, in answer to this, you were to reply: "Yes, it is so. Within this little creature is a factory, that not only makes all you have mentioned, but much more. In this establishment, bones, hair, teeth, nails, flesh, blood, veins, nerves, skin, juices, and water are manufactured; all this is made from milk, and during the first months of the child's life from nothing but milk,"—then your hearer, though he may have the understanding of the most judicious of men, would be dumbfounded, and would beseech you to tell him more about this factory.

You may be certain, he would like to know, how many boilers, cylinders, valves, wires, ladles, oars, pumps, hooks, pins, spokes, and knobs there may be in this factory; more especially would he wish to know, whether the engine of this wonderful establishment be made of steel, wood, cast-iron, silver or gold, or of diamonds.

Now, if you were to tell him, "It contains nothing of the kind. Of all the factories you have seen in your life, there is none that bears any resemblance to this one. And I will tell you furthermore, that it is not even a complete factory, but it is continually developing; it becomes larger and heavier like the child's body itself; moreover, the factory does not consist of iron or steel, nor of gold or diamonds, but it reproduces itself at every moment; it does so merely from the milk that the child drinks,"—then, to be sure, your hearer would begin to doubt his own senses; he would exclaim: "What is the intellect of the intelligent, the judgment of the judicious, what is the wisdom of the wise, when compared to a little of the mother's milk?"

And yet, you are well aware, my friendly reader, that mother's milk is, after all, nothing but milk; and that milk, again, is nothing but a means of nutrition; and nutrition, in its turn, is nothing but a part of the action of the human body.

May I hope that you will favor me with your attention, while, in a few articles, I speak to you about the nutrition of the human body?

## CHAPTER II.

### MAN THE TRANSFORMED FOOD

BEFORE speaking of the process of nutrition in the human body, we must first obtain a correct idea of what is meant by nutrition.

Why are we obliged to eat?

Of course we know that hunger forces us to do so. But every one is aware also, that above all we must ask, whence hunger arises; that we must first get better acquainted with hunger itself, in order to understand nutrition.

To explain this, however, it is necessary to turn our attention to another thing, no less a miracle than nutrition itself, viz., what in scientific language is called "Exchange of Matter." To all of you it is a well-known fact, that nothing in the human body remains even for a moment in the same state; but that in every part of the body a continued exchange takes place. Air is breathed in and exhaled again; but the air exhaled is different from the air inhaled. By this process an exchange of matter has taken place; new matter has entered the body and waste matter has been thrown out.

This exchange of matter—we shall speak more about it at another opportunity—is a principal necessity for the body and its functions; it consists in the main of an incessant change, by which our body is forced to cast out matter that formed parts of it, and is therefore obliged, in order to compensate for the loss, to take in new matter. Hence there is no exaggeration in the expression, "Man is continually renewing himself;" we indeed lose and receive particles of our body at every moment. People have gone so far as to calculate that it takes seven years for the renewal of the whole body of man, and that after this space, there is not even an atom left of the man as he was seven years before.

The regular exchange of matter, as we have seen, supposes the body to be a barter-place, where people take in at the same ratio they pay out. Since, however, man often pays out involuntarily and suffers so many losses—by the mere process of breathing he ejects matter which he must replace afterwards—this exchange of matter is the cause of the body's possessing the feeling of want. The body has paid out and receives nothing in return; this feeling of want is what we call "Hunger." It forces us to absorb as much as we have paid out.

Nutrition, consequently, is the continual replacing of continual losses. It is the wonderful transformation of food into the materials composing the human body.

When looking at our fellow-men, however, we must not think, that they are merely beings that have eaten food; but rather that they themselves, viz., their skin, hair, bones, brain, flesh, blood, nails, and teeth, are nothing but their own food, consumed and transformed.

## CHAPTER III.

### WHAT STRANGE FOOD WE EAT.

MAN, according to what has preceded, is nothing but transformed food.

This idea may frighten us; it may be terrible to our hearts; but let us frankly confess, it is a true one! Man consists only of such substances as he has consumed; he is, in fact, nothing but the food he has eaten; he is food in the shape of a living being.

A child is said to live on his mother's milk; but what else does this mean than: "It is mother's milk, that has become alive by having been changed into head, body, hands, feet, etc., etc."

Indeed, it may sound strange, yet it is quite correct: This mother's milk in the shape of a human being consumes again new mother's milk, and, by respiration, by evaporation and secretion of matter, casts out the used-up milk.

This being so, it will now appear evident to every one, that by a profound chemical knowledge of our daily food, we may readily learn to know the chemical components of man, and *vice versâ;* knowing the substances of which man is made, it is easy for us to determine, what kind of food he must take, in order to continually renew his body.

Since the mother's milk is the simplest and most natural food for the child, let us consider it according to its importance. We shall then have a stepping-stone towards the knowledge of the food of adults and its effects. The mother's milk contains all the elements, with which the human body can renew itself; should there be but one of those elements wanting in it, the child would inevitably perish.

If, for example, milk did not contain calcareous earth, the consequence would be, that the bones of the child would, soon after its birth, neither grow nor increase in number, but they would fast diminish, and the child would die in consequence of this. The attempt was once made to feed animals on articles without calcareous parts, when, strange to behold, they all grew fat, but very weak in their bones, and finally broke down.

If milk contained no phosphorus, not only would the bones and teeth suffer from the want of it, but even the completion of the child's brain could not properly take place, and the child could not replace the quantity of brain which it emits and loses every moment by breathing.

If there were no iron in the mother's milk, the child would die from the green-sickness, a malady which, by the way, is dangerous also for grown people, and which is cured by medicines containing plenty of iron.

If there were no sulphur in it, the child's bile could not develop; the bile, as every one knows, has an important function in the human body.

These are but accessory elements of the mother's milk, elements which usually are not looked upon as articles of food; for who is aware that he must eat, and actually does eat daily, phosphorus, iron, calcareous earth, and sulphur? And not only these; there are a great many other articles, such as magnesia, chlorine, and fluor, that we eat without being aware of it; moreover, our proper food consists also of three gases: nitrogen, oxygen, and hydrogen; and of a solid substance called "carbon," which is no less and no more than pure coal.

All these, my friendly readers, are contained in milk—all these are the elements which in truth constitute the human body. Perhaps some persons believe that there is nothing easier than to procure proper food. It would only be necessary to take a certain quantity of carbon, hydrogen, oxygen, and nitrogen; a little bit of potassium, natron, calcium, and magnesia; to mix a small piece of iron, sulphur, phosphorus, chlorine, and fluor, and take this mixture by the spoon at regular intervals, in order to give the body the necessary aliments. This, however, would be a mistake, for which the perpetrator would pay with his life.

Although it is true that these substances form the proper and most important constituents of our daily food; yet, in order to enjoy the desired result, we must not partake of them in their primary forms; they can actually feed our body only when they are combined together in a peculiar, wondrous manner.

In the next chapter it may be seen how nature first must combine these substances before they are presented to us as proper food; and it will also be seen, that we receive them sometimes in altogether different forms and combinations; for example, in the mother's milk, when we eat the above-named elements in the forms of caseine (cheese), butyrine (butter), sugar of milk, salt, and water.

These latter names have a more savory sound, have they not?

# CHAPTER IV.

## HOW NATURE PREPARES OUR FOOD.

In the preceding article it was stated, that the food of the child which lives on mother's milk, consists in its primary elements of peculiar substances. These are principally oxygen, hydrogen, and nitrogen; three gases to which may be added a large quantity of carbon, or, what is the same, coal. Besides this wondrous mixture of air and coal, the mother's milk contains still other elements, but in a smaller proportion. In every-day life many of them are unfamiliar; for example, natron, calcium, magnesia, chlorine, and fluor; the others, however, are known to every one; viz., iron, sulphur, and phosphorus. All these strange ingredients nature has carefully transformed into milk. For in their primary state, and even in various chemical combinations that may be produced artificially, they would be little adapted for the purpose. It is therefore essentially necessary that nature herself should make them ready for us. This she does by letting them pass first into the vegetable state, and changing them afterwards into new forms.

The plant feeds on primary chemical elements; or, to state it more correctly, the plant is nothing but transformed primary elements! Not before the transformation of these elements into plants are the elements adapted for food for animals and men.

Moreover, all that man eats must first have been in the vegetable state. Now, it is true that man also eats the flesh, fat, and eggs of animals; but whence have the animals meat and eggs? Only from the plants they consume.

There is a remarkable succession of transformations in nature. The primary elements nourish the plant; the plant nourishes the animal; and both, plant and animal, form the nourishment of man.

Even the mother's milk, the simplest and most natural food of the child, owes its existence only to the fact that the mother has eaten vegetable and animal matter. This food, prepared for the mother by nature, has been changed into the body of the same; and partly, also, it has become the milk destined to nourish the child.

Hence it is evident that mother's milk consists of oxygen, nitrogen, hydrogen, carbon, and a small portion of other chemical primary elements. But these substances when appearing in the shape of milk, are combined in such a manner as to form ready-made food; as such they constitute, as stated above, caseine, butyrine, sugar of milk, salt, and water.

The next questions are: "What do these elements of food perform when in the child's body? What becomes of these substances after they have been eaten by the child? How are they changed during the time of their stay in the body? And in what condition do they leave the child's body, and how do they force him to desire food again?"

These questions properly belong to the chapter on "Nutrition," where they will be answered in their turn. Afterwards, we must be permitted to turn our attention to a further question, viz., "What articles of food are the most advantageous to man from the time he is weaned. or the time, he takes from among vegetable and animal matter the same substances for food, that are contained in the mother's milk?"

In order to arrive at the answers to all these questions, we were obliged to first prepare the ground a little This was a gain on our part, for now we shall attain the end in a shorter time than would have been possible otherwise. We trust that we may give our reader a correct idea of the subject, if he will but come to our aid with his most earnest attention and reflection ; these are needed here the more, as we have to treat a difficult subject in a very short space.

## CHAPTER V.

### WHAT BECOMES OF THE MOTHER'S MILK AFTER IT HAS ENTERED THE BODY OF THE CHILD?

When the child has freed itself from the body of its mother, it consists of blood, flesh, and bones, which heretofore were formed and nourished by the blood of the mother.

As soon, however, as the child is born, it ceases to be nourished in this manner. It ceases, also, to secrete through its mother, substances which are useless to it. The child now begins to breathe for itself, and by its breath secretes carbon in the form of carbonic acid. Its skin begins to perspire, and secretes chiefly hydrogen and oxygen in the shape of water or vapor; by the urine, finally, it secretes nitrogen. These substances—carbon, hydrogen, oxygen, nitrogen—before their secretion, constituted vital parts of the child's body; now, however, they are wasted, and for this reason must be thrown off.

It is evident that the child wants compensation for this loss. This is given by the mother's milk; for it contains chiefly these same substances.

But how is this effected?

The milk passes from the child's mouth through the gullet into the stomach. While yet in the mouth, the milk is mixed with a certain liquid called saliva. This saliva possesses the quality of preparing the milk for the necessary change which will take place, when it reaches the child's stomach. The principal work, however, is carried on in the stomach itself. Its sides secrete a liquid called "gastric juice," whose business it is, to transform into a pulp milk, and also solid food, provided the latter be well masticated and moistened.

Science has taught us to prepare gastric juice artificially. The process of digestion, that is, the transformation of solid food—the crust of bread, meat, etc.—into a pulp, may nowadays be observed in a glass filled with warm, artificial, gastric juice.

After the digestion is completed, the lower opening of the stomach, which leads into the duodenum, and which, during the process of digestion, was closed by a muscle, opens itself. The pulp, now called "chyme," flows into the continuation of the stomach—the "alimentary canal" or "duodenum." This is but a long bag with many folds and windings.

The chyme is here mixed again with a liquid called "intestinal juice;" it has the quality of continuing digestion until the chyme separates into two parts; one of them, a milky fluid called "chyle," contains the substance which feeds the body. The other is the solid parts not adapted to nutrition; they are thrown out by the lower opening of the "rectum."

But how is this nutritive part, the chyle, conveyed into the various parts of the body?

The intestinal canal is filled with extremely small vessels called "lacteal absorbents." These vessels absorb the chyle. This absorption, on account of the great length of the intestinal canal—in adults it is nearly thirty feet long—is, in a healthy body, accomplished very thoroughly The real nutriment for the body is now contained in the lacteal absorbents, an infinite number of small tubes.

All these small vessels, however, converge towards the lower part of the spinal column, and uniting, form a vessel which ascends into the chest; here it empties into a large blood-vessel, the blood of which is on its way to the heart.

Thrown out of the heart in another direction, the blood is pushed through the whole body.

Thus the food, after having been transformed into a juice very similar to the blood, joins the blood after a circuitous journey, and is finally mixed with, or, more properly, changed into, blood.

# CHAPTER VI.

### HOW THE BLOOD BECOMES THE VITAL PART OF THE BODY.

ONE would be well justified in calling the blood "man's body in a liquid state." For the blood is destined to become the living solid body of man.

People were astonished, when Liebig, the great naturalist, called blood the "liquid flesh;" we are correct even in going further and calling the blood "man's body in a liquid state." From blood are prepared not only muscles and flesh, but also bones, brain, fat, teeth, eyes, veins, cartilages, nerves, tendons, and even hair.

It is utterly wrong for anybody to suppose, that the constituents of all these parts are dissolved in the blood, say as sugar is dissolved in water. By no means. Water is something quite different from the sugar dissolved in it; while the blood is itself the material from which all the solid parts of the body are formed.

The blood is received into the heart, and the heart, like a pump, forces it into the lungs. There it absorbs in a remarkable manner the oxygen of the air which comes into the lungs by breathing. This blood, saturated now with oxygen, is then recalled to another part of the heart by an expansive movement of that organ.

This part of the heart contracts again and impels the oxygenated blood into the whole body by means of arteries, which branch out more and more, and become smaller and smaller, until at last they are no longer visible to the naked eye. In this manner the blood penetrates all parts of the body, and returns to the heart by means of similar thread-like veins, which gradually join and form larger veins. Having reached the heart, it is again forced into the lungs, and absorbs there more oxygen, returns to the heart, and is again circulated through the whole system.

During this double circulation of the blood from the heart to the lungs and back, and then from the heart to all parts of the body and back again—during all this, the change of particles, so remarkable in itself, is constantly going on: the exchange by which the useless and wasted matter are secreted and new substances distributed. This fact is wonderful, and its cause not yet fully explained by science; but so much is certain, that the blood when being conveyed to all parts of the human body, deposits whatever at the time may be needed there for the renewal of that part.

Thus the blood that has been formed in the child from the mother's milk, contains phosphorus, oxygen, and calcium. These substances, during the circulation of the blood, are deposited in the bones, and form "phosphate of lime," the principal element in the bone. In the same manner fluor and calcium are given to the teeth. The muscles, or flesh, also receive their ingredients from the blood; so do the nerves, veins, membranes, brain, and nails; also the inner organs, such as the heart, lungs, liver, kidneys, intestines, and stomach.

They all, however, in return give to the blood their waste particles, which it carries to that part of the human body where they may be secreted.

If any member of the body is so bound, that the blood cannot circulate, it must decay; for the life of the body consists in its constant change and transformation, in the continual exchange of fresh substances for waste ones. But this vital exchange is only kept up by the constant circulation of the blood, which, while it decreases by being transformed into vital parts of the body, is always formed anew by our daily food.

Food is therefore very justly called "Means of Existence," and the blood may rightly be called the "Juice of Life."

## CHAPTER VII.

### CIRCULATION OF MATTER.

Thus we have seen that the human body is vital blood, transformed and solidified. Now, blood is food transformed; food consists of primary elements prepared and changed by nature; hence, man himself is primary matter transformed and vivified.

But the human race being thousands and thousands of years old, and there being upon the earth besides man the whole of the animal kingdom, developing, preserving, and nourishing itself bodily like man; the question arises: Whence do they all come, these primary elements that are obliged forever to undergo transformation before they can become animated vital matter? Do these primary elements not incessantly decrease during the long process of their being changed into plants and consumed by man and animal, in order to form human and animal bodies afterwards?

The answer to this interesting question has been given already. The human body is not framed or created anew at every moment by food; but it is at every moment, that small particles of the human body die. These particles are returned to mother earth from which they sprang, thus going back to the primary elements.

It is not only those who are dead, that render to the earth what belongs to her, that return to nature what she gave them; but in a far greater degree it is the living, that pay their debt to nature.

Man's body is not his own; nature has lent it to him but for a short term of service; then nature wrests her loan back from him. Thus must man, spite all his pride, accept her never-ceasing offer; daily he must borrow and daily he must repay in part, until the moment comes when he borrows for the last time, the moment he expires; and dying he leaves it to those around his bedside, to pay his last debt to earth.

Is it not wonderful? His own blood is the messenger that daily carries new loans to him, and, in the shape of transformed food, of transformed elements of nature, equips his body. But his own blood is at the same time also his cashier, who, having rendered him service, takes the loan away, by secreting from the body elements that are thus returned to nature.

With every revolution of the blood the body is supplied with transformed food, which is immediately changed into vital parts of the body; with every return of the blood waste matter is carried off and deposited, where it may be thrown out.

The blood carries waste matter to the kidneys that they may send out of the body, in the shape of urine, waste nitrogen, mixed with a part of the phosphate of lime, that served to form bones and teeth, but is now useless. The blood, besides, secretes perspiration through the skin. This is a liquid containing water, hence oxygen and hydrogen; but is moreover mixed with various other waste substances of the body, as for example, carbonic acid, nitrogen and fat. Chiefly, however, the blood is employed in carrying waste carbon to the lungs, so that they may, by the process of respiration, exhale carbonic acid, a gas which would prove of deadly effect if remaining in the lungs too long, or if inhaled.

The quantity of man's secretion per day is by no means small. It amounts to the fourteenth part of his own weight; nay, more—the weight of his perspiration alone, secreted partly by evaporation in the shape of gas, partly as a liquid in drops, amounts during twenty-four hours to nearly two pounds.

Secreted substances have lost all the qualities of transformed and vital matter. They return to the primary elements and serve as food principally to plants, which before had offered those very same substances as food to man.

It is in this manner that the great circulation of matter in nature takes place. From the lifeless primary elements to the plant; from the plant, in the shape of food, to animal and man; from these, as waste substances, back again to the primary elements, there to begin anew a circulation, by means of which inanimate elements are reanimated, and vital elements made lifeless again; that is, life changed again into death.

And it is in this circulation that our "Nutrition," or, more precisely, the "Change of Matter in Man," consists, an important link in the life-preserving chain of nature.

## CHAPTER VIII.

### FOOD.

From what has been said, it must appear evident that only such dishes make good food as contain the same constituents as the blood.

To have these constituents, food must contain salt, fat, and sugar; all these ingredients must, of course, be in a certain proportion.

That water is essential for the support and renewal of the body is clear to every one. The flesh we eat, contains nearly eighty per cent. of water, and yet a man must die, if he were to eat nothing but meat and to have no water, for the reason that the eighty per cent. of water he takes in would by no means be sufficient to form all the liquids necessary for the human body.

The albumen that we eat, forms in the blood chiefly the substances composing the muscular part of the flesh. But it is an error to suppose, that therefore it is absolutely necessary to eat eggs—the white of an egg is nearly pure albumen—because the caseine (cheese) contains precisely the same ingredients as the albumen; for we have seen before, and our readers are doubtless aware of it, that the mother's milk contains caseine, while it is entirely free of albumen. Hence, he who eates plenty of caseine, as do shepherds in Switzerland, for example, scarcely needs any meat. But besides caseine there is another element, viz., the vegetable albumen called gluten, which contains albuminous matter; so do all glutinous plants. Peas, beans, and lentils in particular form food productive of flesh.

The salts that must be given to the blood, do not only consist in the common kitchen-salt. By the expression "Salts" are meant various combinations of substances which are usually not considered articles of food, for example, the combinations of phosphorus, iron, etc., but are not visible to the eye. They help to form bones, teeth, nails, cartilages, and hair.

The fat which we take, appears to many people to be a very important part of our food, and they believe that by eating much fat, one may become fat. But this is not correct. Ferocious animals that live only on meat and fat, do not get fat; while herbivorous animals fatten excessively, if provided with good mast, consisting of course but of plants. Yet fat is, for all this, by no means superfluous to our body. Man needs it, because it is the fat which chiefly supports his respiration. But the fat that is needed for the body, is formed by man himself; so that but little of it need be eaten, and that little only for the purpose of helping to form new fat from sugar.

It is therefore best to consider fat and sugar as food belonging together; for the fat is formed in the body from sugar, and the small quantity of fat which we take daily is only to promote the transformation of sugar into fat.

But let no one believe that one must needs actually eat sugar; no, every food that contains starch supplies the place of sugar very well, as starch is changed, when in the body, first to sugar and then to fat. The potato contains starch and serves its purpose well; it is necessary, however, to put butter with it in order that the starch and sugar formed from the potato in the stomach, may be easily converted into fat.

An excellent article of food is bread, for it contains nearly all the elements of nutrition. It contains vegetable albumen, and therefore is converted into flesh. It has nearly all the salts that are essential to the body; moreover, it contains starch from which fat is produced. Therefore, by the mere addition of a little butter in order to make the formation of fat easier, and by drinking water besides, the human body is able to exist. On the other hand, the potato, if taken alone, is an insufficient means of nutrition. Neither would meat or albumen, if taken alone, be able to preserve life.

Various experiments have been tried with animals, and a great deal of information about the best means of feeding the body has been collected. In order to investigate the effect of the nutritive qualities of food, inquiries have been made especially at military establishments, such as barracks, etc.

# CHAPTER IX.

## ABOUT NOURISHMENT.

In obedience to the demands of modern science, numerous experiments about nutrition have been made, in regard to digestion as well as to the effects of hunger and of various elements of food.

As to digestion, the most excellent observations were made on men afflicted with a fistula in the abdomen, that is, a wound penetrating to the stomach. By means of this wound, it was ascertained very minutely how long it took to digest food, and what kind of transformation it underwent. From this and other experiments it appeared, that the time for digestion, though varying greatly with the various articles of food, lasts from one and one-half to five and one-half hours. Those most quickly digested are: soft sweet apples, beaten eggs, and cooked brain. To digest boiled milk, raw eggs, soft sour apples, roasted beef, liver, two hours were required. Cooked spinal marrow, raw cabbage, fresh milk, roasted beef, oysters, soft-boiled eggs, and raw ham, took nearly three hours. Wheat bread, old cheese, potatoes were digested in nearly three and one-half hours; pork, boiled cabbage, lamb's fat, not before five hours.

The experiments about the effects produced by hunger were tried only on animals. The results were that during the state of starvation three-fourths of the blood disappeared; the fat was almost entirely consumed; the flesh disappeared one-half; even the skin diminished one-third, and the bones lost about one-sixth of their weight. The least decrease was found to be in the nerves, a striking proof that nerves possess a great power of self-preservation, provided there be but a minimum of matter to feed them. From numerous experiments the conclusion was drawn, that an adult weighing about one hundred and thirty pounds must die if he were to lose, say fifty pounds, by starvation.

With regard to the effects of the various articles of food, experiments applied to dogs have shown that they can live on bones for a long time; but that they die if fed on sugar only, and when examined after death, no trace of any fat is to be found.

Animals fed on substances that contained no phosphorus and lime became fat; but they died for want of the proper nourishment for their bones. Animals died also when nourished only with pure albumen or pure caseine. The most remarkable fact in this connection is, that they perished in the same length of time in which they would have died, *if they had taken no food whatever.*

Experiments tried on man have shown that it is injurious to eat *uniform* food. A constant change in our food is extremely nourishing and healthy. This is an experience made in prisons and barracks; changes of food are made there every day during the week, so that each day they have a different dinner. Once, a physician in England wished to try the effects of uniform food on himself. He took nothing but bread and water for forty-five days; in consequence of this he decreased eight pounds. Then he ate for four weeks but bread and sugar, then bread and oil three weeks; but finally he succumbed under his experiments, and died, after having experimented thus for eight months.

We must not, therefore, call it daintiness when we feel an appetite for more variety of food, or if we soon get tired of uniform meals: a constant change in this respect is necessary. Experiments have shown that rabbits continue their health, if alternately they receive one day potatoes, the next day barley; but if they receive exclusively potatoes or barley, they soon die.

In conclusion, we will mention a few articles of food and their qualities. Among grains, wheat is known to be the most nutritive, and wheat bread and meat taken together is always good, wholesome food. Rice produces fat, but if taken by itself, it is not worth much, since it is nourishing only if eaten with butter, or fat, and a little meat. Potato is a cheap, and yet an expensive food; for it contains very little nutriment. In order to be of benefit it must be eaten in great quantity; besides, it is necessary to season it with salt, butter, or fat, as otherwise it would be totally useless. A good diet is peas, beans, and lentils; but their hulls are indigestible, and must be removed.

In general, beverages are not counted among articles of food; and kitchen-salt is commonly believed to be but a matter of taste; but this is a great mistake. Coffee and tea, too, are nourishing in their way; good beer is equal to half a dinner, and as to salt, a frequent relish of the same is an excellent means of nutrition.

Cheap coffee, cheap beer, and cheap salt are therefore a great benefit to the people.

# PART IV.

# LIGHT AND DISTANCE.

# CHAPTER I.

## SOMETHING ABOUT ILLUMINATION.

From time to time we hear of plans to illuminate whole cities by a great light from a single point. The credulity of the newspaper public about affairs belonging to Physics is so great, that we are not surprised if such plans are spoken of as practicable; though, indeed, one needs but cast a glance of reflection on them, to be at once convinced of their impracticability.

The impracticability does not consist so much in this, that no such intense light can be made artificially, as in the circumstance that the illuminating power of light decreases enormously as we recede from it.

In order to explain this to our readers, let us suppose that on some high point in New York city, say Trinity-church steeple, an intensely brilliant light be placed, as bright as can be produced by gases or electricity. We shall see, presently, how the remoter streets in New York would be illuminated.

For the sake of clearness, let us imagine for a moment, that at a square's distance from Trinity church there is a street, intersecting Broadway at right angles. We will call it "A" street. At a square's distance from "A" street let us imagine another street running parallel to it, which we will call "B" street; and again, at a square's distance, a street parallel to "B" street, called "C" street; thus let us imagine seven streets in all—from "A" to "G"—running parallel, each at a square's distance from the other, and intersecting Broadway at right angles. Besides this, let us suppose there is a street called "X" street, running parallel with Broadway and at a square's distance from it; then we shall have seven squares, which are to be illuminated by one great light.

It is well known that light decreases in intensity the further we recede from it; but this intensity decreases in a peculiar proportion. In order to understand this proportion we must pause a moment, for it is something not easily comprehended. We hope, however, to present it in such a shape, that the attentive reader will find no difficulty in grasping a great law of nature, which, moreover, is of the greatest moment for a multitude of cases.

Physics teach us, by calculation and experiments, the following:

If a light illuminates a certain space, its intensity at twice the distance is not twice as feeble, but two times two, equal four times, as feeble. At three times the distance it does not shine three times as feeble, but three times three, that is nine times. In scientific language this is expressed thus: "The intensity of light decreases in the ratio of the square of the distance from its source."

Let us now try to apply this to our example.

We will take it for granted that the great light on Trinity steeple shines so bright, that one is just able to read these pages at a square's distance, viz., on "A" street.

On "B" street it will be much darker than on "A" street; it will be precisely four times darker, because "B" street is twice the distance from Trinity church, and $2 \times 2 = 4$. Hence, if we wish to read this on "B" street, our letters must cover four times the space they do now.

"C" street is three times as far from the light as "A" street; hence it will be nine times darker there, for $3 \times 3 = 9$. This page in order to be readable there, would then have to cover nine times the space it occupies now.

The next street, being four times as remote from the

light as "A" street, our letters, according to the rule given above, would have to cover sixteen times the present space, for it is sixteen times darker there than on "A" street.

"E" street, which lies at five times the distance from the light, will be twenty-five times darker, for $5 \times 5 = 25$. "F" street, which is six times the distance, we shall find thirty-six times darker; and, lastly, "G" street, seven times the distance from the light, will be forty-nine times darker than "A" street, because $7 \times 7 = 49$. The letters of a piece of writing, in order to be legible there, must cover forty-nine times the surface that our letters cover now.

But the reader will exclaim: "This evil can be remedied. We need but place forty-nine lights on Trinity steeple; there will then be sufficient light on "G" street for any newspaper to be read." Our friend will easily perceive, however, that it is more judicious to distribute forty-nine lights in different places on Broadway, than to put them all on one spot.

This is sufficient to convince any one that we may be able to illuminate large public places with *one* light, but not the streets of a city, and still less whole cities.

## CHAPTER II.

### ILLUMINATION OF THE PLANETS BY THE SUN.

It was demonstrated above, that it is impossible to illuminate large distances by a single light. Yet we must acknowledge that nature herself does this, and that the sun is the only light which shines throughout the solar system; for the light which is seen in the planets is but light received and reflected from the sun.

This is sufficient reason for us to believe, that there are not on every planet creatures as we see them on our earth; but that, on the contrary, each celestial body may be inhabited by creatures organized according to the distance of the planet from the sun; that is, adapted to the degree of light produced there by the sun.

For the natural sciences teach us, that solar light is subject to the same laws as our artificial light: it decreases as the distance increases. The planets more remote from the sun are illuminated less than those nearer to it. The ratio in which this light decreases, is precisely the same as that of the terrestrial light illustrated above, viz., according to the square of the distance. In other words, when the distance is double, the intensity of the light is one-fourth as great; when three times, one-ninth as great; when four times more remote, one-sixteenth as strong, etc.; in short, at every distance as much weaker as the distance multiplied by itself.

Presently we shall see that the planets are illuminated in inverse proportion to their distance from the sun. From this alone we come to the conclusion, that on every planet the living beings must necessarily be differently constituted.

The name of the planet nearest to the sun is Mercury. It is about two and a half times nearer to the sun than our earth, therefore it receives nearly seven times as much light. We can scarcely conceive such an intensity of light and all the consequences resulting from it. If instead of one sun we should happen to have three, there is no doubt that we should go blind; but seven suns, that is, seven times the light of our brightest days, we could not endure, even if our eyes were closed; the more so, as our eyelids, even when firmly closed, do not protect us from the sun's light entirely. This is a proof of our assertion, that the living beings on the planet Mercury must be differently organized from us.

Venus, the third planet, is one and a third times nearer to the sun than we are. The light on that planet, therefore, is nearly twice as bright as on ours. But inasmuch as even this would be unbearable for us, the creatures on this planet must likewise be different from us.

The third planet is the earth we inhabit. The intensity of the sunlight in bright summer days is well known to us from experience, although no one has as yet been successful in measuring its degree as precisely as has been done with heat by the thermometer. It is true that in modern times a certain Mr. Schell, in Berlin, proposed to measure light accurately, in a way that elicited the approbation of naturalists, especially of Alexander von Humboldt. However, the experiments proposed have not yet been properly carried out, though they are very useful to photographists. Therefore we do not know, up to the present time, whether there is any difference in the light of two cloudless summer days; just as little are we able to determine how much the moon's light is weaker than the sun's.

The fourth planet's name is Mars; its distance from the

sun is one and a half times our distance from the sun. There the sun's light is about half as strong as with us. Now, although we often may have days which are half as bright as others, it is yet very doubtful whether we could live on Mars; for light does not act upon our eyes only, but on our whole body and its health. It is likely that the very want of light there would prove fatal to us.

The twenty-four newly discovered planets have days that are nearly six times darker than ours. The daylight on these planets is probably as it was with us during the great eclipse of the sun in July, 1851. This light was very interesting for a few minutes, but if it were to continue it would certainly make us melancholy.

Far worse yet fare the remoter planets. On the planet Jupiter it is as much as thirty times darker than with us. On Saturn, eighty times. On Uranus, even three hundred times; and upon the last of the planets, Neptune, discovered in 1845, light is nine hundred times more feeble than upon our globe.

Although it is true that all of the remoter planets have many moons or satellites, yet it must not be forgotten that the moons themselves are but very feebly illuminated; that their light benefits during the night only, and even then only lovers and night revellers.

# PART V.

# THE WONDERS OF ASTRONOMY.

# CHAPTER I.

## A WONDERFUL DISCOVERY.

Many people are greatly surprised, that when a new planet is discovered—and within late years this has been frequently the case—astronomers should be able to determine a few days afterwards its distance from the sun, together with the number of years necessary for its orbit. "How is it possible," they ask, "to survey a new guest after such a short acquaintance so accurately, as to foretell his path, nay, even the time of his course?"

Nevertheless it is true that this can be done, and certainly no stage-coach nor locomotive can announce the hour and minute of its arrival with as much accuracy as the astronomer can foretell the arrival of a celestial body, though he may have observed it but a short time.

More yet is done sometimes. In 1846, a naturalist in Paris, Leverrier by name, found out, without looking in the sky, without making observations with the telescope, simply by dint of calculation, that there must exist a planet at a distance from us of 2,862 millions of miles; that this planet takes 60,238 days and 11 hours to move round the sun; that it is 24½ heavier than our earth, and that it must be found at a given time at a given place in the sky; provided, of course, the quality of the telescope be such as to enable it to be seen.

Leverrier communicated all this to the Academy of Sciences in Paris. The Academy did not by any means say, "The man is insane; how can he know what is going on 2,862 millions of miles from us; he does not even know what kind of weather we shall have to-morrow!" Neither did they say, "This man wishes to sport with us, for he maintains things that no one can prove to be false!" Nor, "The man is a swindler, for he very likely has seen the planet accidentally, and pretends now that he discovered it by his learning." No, nothing of the kind; on the contrary, his communication was received with the proper regard for its importance; Leverrier was well known as a great naturalist.

Having thus learned how he made the discovery, the members of the Academy felt convinced that there were good reasons to believe his assertions to be true.

Complete success crowned his efforts.

He made the announcement to the Academy in January, 1846; on the 31st of August he sent in further reports about the planet, which he had not seen as yet. The surprise and astonishment on the part of scientific men can scarcely be imagined, while on the part of the uneducated there were but smiles and incredulity.

On the 23d of September, Mr. Galle—now Director of the Breslau Observatory, at that time Assistant in that of Berlin, a gentleman who had distinguished himself before by successful observations and discoveries, received a letter from Leverrier, requesting him to watch for the new planet at a place designated in the heavens. Though other cities at that time possessed better telescopes than Berlin, this city was chosen because of its favorable situation for observations.

That same evening Galle directed his telescope to that spot in the sky indicated by Leverrier, and, at an exceedingly small distance from it, actually discovered the planet.

This discovery of Leverrier is very justly called the greatest triumph that ever crowned a scientific inquiry. Indeed, nothing of the kind had ever transpired before; our century may well be proud of it. But, my friendly reader, he who lives in this age without having any idea whatever of the way in which such discoveries are made—he does not deserve to be called a contemporary of this age.

We will not try to make an astronomer out of you; we merely wish to explain to you the miracle of this discovery.

# CHAPTER II.

## MAIN SUPPORT OF LEVERRIER'S DISCOVERY.

WHEN Leverrier was working at his great discovery he did not strike out a new path in science; he was sup ported by a great law of nature, the base of all astronom ical knowledge. It is the law of gravitation, discovered by Sir Isaac Newton.

Those of our readers who have fully understood what we said before (page 50) about light, will now easily comprehend, what we are going to say about the force of gravity.

Every heavenly body is endowed with the power of attraction; that is, it attracts every other body in the same manner that a magnet attracts iron. If the celestial bodies, or, to speak only of one class, if all the planets were at rest, that is, without motion, they would, on account of the great attractive power of the sun, rapidly approach it, and finally unite with it and form one body.

That this does not take place, may be ascribed solely to the fact that all planets have their own motion. This motion, combined with the attractive force of the sun, causes them to move in circles around it.

This may be illustrated by the following: Suppose a strong magnet to lie in the centre of a table. Now, suppose some one to place an iron ball on the table; then will this ball run straightway towards the magnet. But if some one were to roll the ball so that it should pass the magnet, it would at first run in a straight line, but the magnet attracting it at every moment of time, the ball would be compelled to deviate from its straight course and would begin to circulate round the magnet.

We see that this circular motion round the magnet springs from two forces: first, from the hand that starts the ball in a straight line; and secondly, from the attraction of the magnet, which at every moment draws the ball towards itself.

Newton, the greatest natural philosopher of all times, who lived in England two hundred years ago, proved, that all the orbits round the sun, as described by the planets, are caused by two such forces; by the motion of the planets peculiar to themselves, which, if not interfered with, would make them fly through space in a straight line; and by the attractive force of the sun, which is continually disturbing that straight course, thus forcing the planets to move in circles around him.

But Newton has discovered more than this. He succeeded in proving that, knowing the time of a planet's revolution around the sun, we can determine precisely with what force the attractive power of the sun affects it. For if the sun's attractive power is strong, the planet will revolve very quickly; if weak, it will move slowly.

Were the sun, for example, all of a sudden to lose a portion of his attractive force, the consequence would be that the earth must revolve around him more slowly. Our year, which now has three hundred and sixty-five days, would then have a much greater number of days.

Newton has also shown—and this is for us the main thing—that the attractive force of the sun is strong in his close proximity, but that it diminishes as the distance from him increases. In other words, the remoter planets are attracted by the sun with less force than those nearer to him. The attractive force decreases with the distance in the same proportion as light, which, we saw a little while ago, decreases in intensity as the square of the distance increases. This means, that a planet at a distance from the sun twice as great as that of the earth, is attracted with only one-fourth the force; one that is three times the distance, with one-ninth of the force, etc.

This great law pervades all nature. It is the basis of the science of astronomy, and was the main support of Leverrier's discovery.

# CHAPTER III.

## THE GREAT DISCOVERY.

PERHAPS the question presents itself to the thinking reader: If it be true that the heavenly bodies attract each other, why do not the planets attract one another in such a manner that they will run round and about each other?

Newton himself proposed this question; he also found the answer. The attractive power of a celestial body depends upon its larger or smaller mass. In our solar system the sun's mass is so much larger than that of any of the planets, that the balance of attractive power is largely in his favor; hence the revolving of the planets around him. If the sun were to disappear suddenly the effect of the attractive influence of the planets upon one another would be tremendous. There can be no doubt that they would all begin to revolve around Jupiter, because that planet has the largest mass. To give some examples in figures,—the sun's mass is 355,499 heavier, while Jupiter's is but 339 times heavier than that of the earth. It is evident that, the sun's mass being more than a thousand times larger than Jupiter's, so long as the sun exists the earth will never revolve around Jupiter.

Yet Jupiter is not without influence upon the earth; and though it is not able to draw it out of its course round the sun, yet it attracts the earth to some extent. Observations and computations have shown us, that the earth's orbit around the sun, owing to the attraction of Jupiter, is somewhat changed, or, as it is called, "disturbed."

As with Jupiter and the earth, so with all the other

planets; their mutual attraction disturb their orbits round the sun. In reality, every planet revolves in an orbit which, without this "disturbance," would be a different one. The computation of these disturbances constitutes a great difficulty in astronomy, and requires the keenest and most energetic studies ever made in science.

Perhaps some of our readers may ask here, whether in course of time these disturbances will become so great as to throw our whole solar system into confusion? Well, the same question was proposed by a great mathematician named Laplace, who lived towards the end of the last century. But he himself answered the question in an immortal work, "The Mechanics of the Heavens." He furnished the proof, that all disturbances last but a certain time; and that the solar system is constructed so that the very attractions by which the disturbances are caused, produce at the end of certain periods a regulation or rectification; so that in the end there is always complete order.

After what has been said, it is evident that if one of the planets were invisible, its presence would still be known to our naturalists, on account of the disturbances it would cause in the orbits of the other planets; unless, perhaps, its mass be so insignificant as to render its power of attraction imperceptible.

And now we may proceed to explain the subject of this chapter.

Up to the year 1846, when Leverrier made his great discovery, it was believed that Uranus was the most distant planet revolving around the sun. Uranus itself was discovered by Sir John Herschel in England in the year 1781. As this planet takes eighty-four years to go round the sun, its complete revolution had not yet been observed in 1846; in spite of this, however, the course of Uranus was calculated and known very precisely, because the attractive force of the sun was known; and all the disturbances that might influence the planet were taken into account.

But notwithstanding all nicety of calculations, the real course of Uranus would not at all agree with the one computed. At that time already, long before Leverrier's discovery, the idea arose that beyond Uranus, in a region where the human eye could, in spite of all telescopes, discover nothing, there must probably exist a planet which changed the course of Uranus. Bessel, a great astronomer, who unfortunately for science died too soon, was already on the point of finding out by computation the unknown disturber. But he died, shortly before Leverrier's discovery. As early even as 1840, Maedler, in the city of Dorpat, in Russia, wrote a fine article on this as yet unseen disturber.

Leverrier, however, began the task and finished it. He computed with an acuteness that was admired by all men of science. He investigated whereabout in the heavens that intruder must be situated, so as to be able to trouble Uranus to such an extent; how fast this disturber itself must move in its orbit, and how large must be its mass.

We live to see the triumph of Leverrier's being able to discover with his *mental* eye, by means of computation only, a planet at a distance of millions of miles from him.

Therefore let us say: Honor science! Honor the men that cultivate it! And all honor to the human intellect which sees farther than the human eye!

# PART VI.

# METEOROLOGY.

# CHAPTER I.

## SOMETHING ABOUT THE WEATHER.

We presume that in a state of unusual bad weather there are many persons, who find occasion to reflect on the nature of weather in general.

A few years ago, we had "green Christmas and white Easter," and spring was of course far behind when Pentecost arrived. We had still cold and rainy days, while the nights were frosty; and, if one might judge from appearances, it seemed that nature had made a mistake, and had not known of our being then in the month of June, which, with us, is usually a delightful month.

The sun alone was right. He rose on the 9th of June of that year precisely at 4 o'clock 30 minutes, as was prescribed to him by the calendar; and set at 7 o'clock 30 minutes, precisely according to orders. At that time the sun was hastening towards summer, he lengthened the days and shortened the nights; but he alone is not capable of governing the weather, and our friends the astronomers, although they are able to calculate the sun's course with more precision than the engineer can the locomotive's, are themselves greatly embarrassed when asked, "What kind of weather shall we have the day after to-morrow?"

It is unpardonable that some of our almanacs, especially those for the farmer, contain prophecies about the weather. We cannot be too indignant against the foolish superstition which this abuse tends to foster. And what is worse, really shameful, is, that those who print such things do not believe in them themselves, but consider them a necessity sanctioned by age and custom, and offer it as such to the credulity of the public.

The subject of this article on the knowledge of weather, is a science, a great branch of the natural sciences; but it is a branch just developing, and therefore has, up to the present time, not yet brought forth any fruit.

It is very likely that at some future day we shall be able to indicate in advance the weather of any given place. But for the present this is impossible; and if from time to time men arise and announce that they can calculate and determine in advance the state of the weather in any given place—pretending to consult the planets, etc.—we take it for granted that they are as unreliable as the weather-prophets of the almanacs.

We said above that the weather might possibly be determined a few days ahead; science is at present almost far enough advanced for it. But there are needed for that purpose grand institutions, which must first be called into life.

If for the proper observation of the weather, stations were erected throughout the extent of our country, at a distance of about seventy miles from each other, and if these stations were connected by a telegraph-wire, managed by a scientific reliable observer; then we might, in the middle portion of our country, be able to determine in advance the state of the weather, though for a short time only.

For the changeableness of the weather depends on the nature and motion of the air, and on the amount of moisture, and the direction of the winds. It is mostly occasioned by currents of air which pass over the earth, producing, wherever they meet, here cold, there heat—here rain, there hail or snow.

Along a part of the coast of the United States electric telegraphs have been established. Vessels receive, at a considerable distance, the news of a storm approaching, together with its velocity and direction. The electric telegraph being quicker than the wind, the vessels receive the news in time to take their directions. Before the storm reaches them, they have been enabled to take precautionary measures for its reception.

This is a great step forward in our new science. But not before the time when such stations shall be established everywhere throughout the land, will meteorology manifest its real importance. For it has, like every other science, firmly established rules, which can easily be calculated and verified; while, on the other hand, allowances must be made for changeable conditions which tend to disturb the rules.

We will now endeavor to introduce to our readers these established rules, and explain the changeable conditions to which we refer.

# CHAPTER II.

## OF THE WEATHER IN SUMMER AND WINTER.

As we have stated above, there exist fixed rules about the weather; these rules are simple and easy to compute. But our computations are often disturbed by a great many circumstances beyond our reach, so much that we are governed more by exceptions than rules.

These latter are based on the position of our earth with regard to the sun. They are, therefore, easy to determine, for astronomy is a science resting on firm pillars; and although nothing in the universe is so far from us as the stars, yet there is nothing in the world so certain as our knowledge of the courses of the constellations and their distances. Many of our readers may be surprised, perhaps, to hear that we know more accurately the distance from the earth to the sun than the distance from New York to Cincinnati. Indeed, astronomical knowledge is the most reliable in the world. No merchant is able to measure a piece of cloth without being mistaken, to say the least, as much as $\frac{1}{300}$ part; while the uncertainty with respect to distances of bodies in the solar system amounts to a great deal less than $\frac{1}{300}$ part.

Our earth turns on its axis once in every twenty-four hours, and goes also round the sun once a year. But the earth's axis is inclined towards the earth's orbit—orbit is the circle which a celestial body describes in its revolution around another—in such a manner as to cause the earth, in its orbit round the sun, to be illuminated for six months on one side, and for six months on the other side of the earth. Hence it happens, that at the north pole there is continual day during six months in the year, after which follows uninterrupted winter for the next six months; in the same way the day on the south pole lasts six months, and the night following the same length of time. In the middle between both poles, however, in the regions around the equator, the day has throughout the year twelve hours; the night, of course, the same; while in the countries between the equator and the poles, the length of day and night is, through the whole year, constantly varying.

We, in the United States, inhabit the northern hemisphere; when, therefore, the time comes that the north pole has day for six months, we in North America, being situated about half-way between the equator and north pole, enjoy long days and short nights. The inhabitants of those countries, however, situated on the southern hemisphere, have at that time short days and long nights. But when the time comes that there is six months' night on the north pole and six months' day on the south pole, then will the inhabitants of the southern hemisphere have long days, and we long nights.

Intimately connected with the length of day and night are our seasons, especially summer and winter; for together with the sun's light heat is also called forth. During our long days, therefore, it is very warm with us, for the sun's rays heat the soil. During our short days we experience cold, because the warming light of the sun does not reach our earth directly. For this reason the northern hemisphere enjoys summer while the southern has winter; and *vice versâ*, when we have mid-winter, people in the other hemisphere are in the midst of summer. When we are snowed up at Christmas, and seek joy and elevation by the cheerful fireside in the brightly-lighted room, we may, perhaps, think of our friends and relatives who have emigrated to Australia, and the question may occur to us, how things may be with them this cold weather, and how they are enjoying the holidays?

Now, would not the uninformed be surprised, if a letter were to arrive from Australia, written at Christmas, telling how the writer enjoyed Christmas in his vine-arbor, where he had sought shelter from the terrible heat of the day, and that he had but late at night gone to his room, and he could scarcely sleep then on account of the heat, and the longing for his former home in the United States, where he could always enjoy cool weather at Christmas.

The uninformed will now learn that Australia lies in the southern hemisphere, while we are in the northern, and that there they live in midst of summer, while we are buried in snow. Nor will he now be surprised when he reads, that it snowed in Australia in the month of August, and that his friend or relative there reposed by the fireside, and read the letter from home by the light of the lamp, at the same hour that we here were taking an afternoon walk in the summer shade.

The heat of summer, however, does not altogether depend upon the length of the day; nor does the cold of winter upon its shortness; but principally on this, that during summer-time the sun at noon stands directly over head; that therefore his vertical rays are enabled to pierce the soil with intense heat; while in winter-time the sun at noon stands nearer to the horizon; his rays fall on the earth obliquely, therefore heating the soil with but feeble power.

We shall presently see that this position of the sun exercises great influence upon the weather.

# CHAPTER III.

## THE CURRENTS OF AIR AND THE WEATHER.

In order to fully understand the conditions of the atmosphere, one must carefully notice the following:

Though the sun produces summer and winter, and although his beams call forth heat, and the absence of heat causes intense cold on the surface of the globe, yet the sun alone does not make what we call "Weather."

If the sun's influence alone were prevalent, there would be no change at all during our seasons; once cold or warm, it would invariably continue to be so, according to the time of the year. The sun, however, produces certain movements in the air; currents of air or winds pour from cold countries into warm ones, and *vice versâ* from warm ones into cold ones. It is this that makes our sky be cloudy or clear; that produces rain and sunshine, snow and hail, refreshing coolness in summer and warmth sometimes in midwinter, as also chilly nights in summer and thaw in winter. In other words, it is more properly the motion of the air, the wind, that produces what we call *weather;* that is, that changeableness from heat to cold, from dryness to moisture, all of which may be comprised in one name, weather.

But whence does the wind arise? It is caused by the influence of the sun's heat upon the air.

The whole earth is enveloped with a misty cover called "air." This air has the peculiar quality of expanding when it becomes heated. If you put a bladder that is filled with air and tied up, into the pipe of a heated stove, the air inside will expand so much as to burst the bladder with a loud report. The warm expanded air is lighter than the cold air, and always ascends in the atmosphere.

Lofty rooms are therefore difficult to heat because the warm air ascends towards the ceiling. In every room it is much cooler near the floor than near the top of the room. This accounts for the singular fact that in winter our feet, though warmly clad in stockings and shoes or boots, feel cold more often than our hands, which are entirely uncovered. If you ascend a ladder in a tolerably cold room, you are surprised at finding it much warmer above than below in the room. The flies take advantage of this in autumn, when they are seen to promenade on the ceiling, because there it is warm as in summer, while near the floor it is cold; owing to the circumstance that warm air, being lighter than cold, ascends.

Precisely the same takes place on the earth. In the hot zone near the equator the sun heats the air continually; hence the air there ascends. But from both the northern and southern hemispheres, cold air is constantly pouring towards the equator in order to fill the vacuum thus produced. This cold air is now heated also and rises, while other cold air rushes in after. By this continued motion of the air towards the equator, however, a vacuum is created also at both poles of the earth; and the heated air of the equator, after having ascended, flows towards these two vacuums. Thus arise the currents in the air; currents which continue the whole year, and cause the cold air to move from the poles to the equator along the surface of the earth; while higher in the atmosphere the heated air flows from the equator back to the poles.

Therefore the air is said to circulate below from the poles to the equator, but above to go back from the equator to the poles.

He who is in the habit of noticing phenomena of

nature, may often have observed something of the kind when opening the window of a room filled with smoke. The smoke escapes above, while below it seems to come back into the room again.

But this is an illusion which has its origin in the fact, that above the warm air of the room goes out of the window, and, of course, takes the smoke with it; below at the window, however, cold air pours in from without, driving the smoke that is below back into the room. The attentive observer may also see how the two currents of air above and below move in contrary directions; while in the middle part they repel each other, and form a kind of eddy which may be clearly perceived by the motion of the smoke.

What takes place on our earth is nothing different from this, and we shall presently see the great influence this has upon our weather.

# CHAPTER IV.

## THE FIRM RULES OF METEOROLOGY.

The air which is continually rising in the hot zones and circulating towards the poles and back again to the equator, is the prime source of the wind. This latter modifies the temperature of the atmosphere; for the cold air from the poles of the earth, in coming to the equator, cools the torrid zone; again, the hot air going from there to the poles heats the colder regions. This accounts for the fact that very often it is not so cold in cold countries as it really would be, were it not for this circulation of the air; and that in hot countries we never find the degree of heat that there would be if the air were continually at rest.

According to what has been said, however, but two different winds would exist on the earth, and these two moving in fixed directions; one sweeping over the earth from the poles to the equator, with us called "North wind," and one from the equator to the icy regions, with us the "South wind."

But we must add here something which considerably modifies this, viz., the revolution of the globe. The earth, it is well known, revolves round its axis from west to east once in twenty-four hours; the atmosphere performs this revolution also.

But since that part of the atmosphere nearest to the equator must move with greater velocity than the part nearer the poles, it may with a little thinking be easily understood, that the air which goes on the surface of the earth from the poles to the equator, passes over ground which moves faster east than the air itself; while, on the contrary, the air coming from the hot zone starts in an eastern direction with the velocity it had on the equator; but, as it it is moving on, it passes over that part of the earth which rotates with less velocity.

This gives rise to what are called the *trade-winds*, so very important to navigation. In our hemisphere the trade-winds come in the lower strata of the air from the north-east; while in the upper strata they move towards north-east, they come from the southwest. On the other hemisphere the trade-winds in the lower strata of the air move in a northwesterly direction; in the upper they move in a southeasterly direction.

From this arise our rules respecting the weather.

The idea that many persons have that wind and weather are two things entirely different, is wrong. Weather is nothing else but a condition of the atmosphere. A cold winter, cold spring, cold summer, and cold autumn, do not mean, as some believe, that the earth, or that part of it on which they live, is colder than usual; for if we dig a hole in the ground, it will be found that neither cold nor warm weather has any influence upon the temperature below the surface of the earth. At the small depth of thirty inches below the surface, no difference can be found between the heat of the day and the cold of the night. In a well sixty feet deep no difference is perceivable between the hottest summer and the coldest winter-day, for below the surface of the earth the differences of temperature do not exist. What we call Weather" is but a state of the atmosphere, and depends solely upon the wind.

It has been stated already that there are fixed rules of weather, or, which is the same thing, that there are laws governing the motion of the winds; but we have added also, that there are a great many causes which disturb these rules, and therefore make any calculations in advance a sheer impossibility.

We have seen that these rules are called forth, 1st, by the course of the sun; 2d, by the circulation of the air from the poles to the equator and back again; and 3d, by the revolution of the earth, causing the trade-winds.

All these various items have been calculated correctly; and, owing to this, we have now a firm basis in meteorology. But in the next article, we shall see what obstacles are put in the way of this new science by other things; and the allowances to be made for these disturbances cannot be easily computed.

## CHAPTER V.

### AIR AND WATER IN THEIR RELATIONS TO WEATHER.

LET us now examine the causes which disturb the regular currents of air, and which render the otherwise computable winds incomputable, thus producing the great irregularities of the weather.

The main cause lies in this, that neither the air nor the earth is everywhere in the same condition.

Every housewife that but once in her life hung up clothes to dry, knows full well that air absorbs moisture when passing over, or through, wet objects. If she wishes to dry her clothes very quickly, she will hang them up where there is much wind. And she is perfectly right in maintaining that the wind dries clothes better than the quiet sunshine.

Whence does this come?

From this: dry air, when coming in contact with wet objects, absorbs the moisture, and by this dries the object somewhat. If there be no wind, the moistened air will remain around the wet object, and the drying goes on very slowly. But so soon as a little wind arises, the moist air is moved away, new dry air constantly takes its place, and coming into contact with the wet article, effects in a very short time the desired result.

Hence, it is not heat alone that causes the clothes to dry; for in winter-time, though it is so cold that the clothes on the line freeze to stiffness, they dry nevertheless, if it be very windy. It is the wind which dries them by allowing fresh air to pass through them continually. For the same reason our housewives open doors and windows after a room has been scoured, so that by a thorough draft of air, the floor may dry quickly; a large fire in the stove or fireplace could not effect it so readily.

From all this we may learn that the air absorbs particles of water. It will now be evident to every one, why water in a tumbler, standing uncovered at the open window for a few days, constantly decreases, until it finally disappears entirely and the tumbler is dry. Where has the water gone? The air drank it off, little by little, until at last the tumbler was emptied.

"But," you will exclaim, "what does the air do with all the water it drinks? The air goes over the whole ocean; over lakes, rivers, brooks, and springs; over woods and fields, and everywhere it takes in particles of water. What becomes of them?"

After being absorbed, the particles of water unite and form clouds; then they fall down in the form of fog, rain, snow, or hail.

Many persons, even highly educated ones, have false ideas about these phenomena of the atmosphere.

Some think a cloud is a kind of bag that contains the rain which is let fall by the cloud. This is entirely false. The clouds are nothing but fogs in the upper regions of the atmosphere; fog itself is nothing but a cloud immediately over ground.

It is easy to obtain a correct idea of the formation of fog and rain; one need but observe for one's self.

He who has ever blown upon his hands in winter-time in order to warm them, will have observed that his hands become moist from his breath. If a window-pane is breathed upon, it is covered by a thin coat of water. What is the cause of this? It arises from the fact that the air we exhale contains water-particles from our blood. We do not see them when it is warm, because they are airy themselves; everybody knows that they become visible so soon as the air turns cool; or that they appear like fog when one is in a cold room in winter; that they form drops when you breathe upon cold objects; that they freeze and become snow; nay, that in severe cold weather, after a long walk out-doors, they even cling to one's moustache like icicles.

This may illustrate, that these particles of water are invisible in the warm air, but that when the air is colder they appear as fog; when still colder, as drops of rain; and in very cold weather they turn to snow, while in severe cold they freeze and form ice.

# CHAPTER VI.

## FOG, CLOUDS, RAIN, AND SNOW.

THE air imbibes particles of water from all parts of the earth; and thus charged with water it is the same and operates the same as our breath.

So soon as a stratum of air that contains water-particles, meets with a colder stratum, these airy particles of water immediately flow together to form fog. But fog, as has been said, is nothing but a cloud. He who has travelled in mountainous countries, has often noticed this. From the valley it often appears that the top of a high mountain is wrapped in clouds; and his curiosity may be excited to ascend the mountain in order to examine these clouds. But when he arrives there, he sees nothing whatever either before or behind him but fog, which most assuredly he has often seen before without so much trouble. The ignorant person who believes that a cloud is something else than fog, and who fancies that the clouds which he saw from below have disappeared during his ascent, leaving but a mist behind, will be no little amazed when he has arrived at the foot of the mountain again, to see the cloud above as before, and to perceive that he actually walked among the clouds.

Hence it is understood now, that the particles of water in the air form fog, or, which is the same, clouds, so soon as they come into a colder stratum. But the cloud is not rain as yet; the change into rain will depend upon circumstances that may be easily guessed. If a warmer and dryer stratum passes over the one containing the newly formed clouds, then this warmer stratum will absorb the water-particles of the other. The moist air fares like the wet clothes we spoke of; the warm dry air absorbs its particles of water. But if a colder stratum of air approaches the stratum containing clouds, then the water-particles of the latter are condensed; the cloud becomes small drops of water; these drops are too heavy to be supported in the air, and they fall down as *rain.*

During its descent, the drop of rain is steadily increased by the water-particles of the air through which it passes. Thus it happens, that rain often arrives at the earth in the form of large drops of water, while when yet in the air and beginning to fall, it consisted of tiny drops. It is well known that the rain-drops on the roof are smaller than those that fall on the street. The difference is so great, that on the roof of the royal castle in Berlin, Prussia, there falls four and a half inches less rain during the year than on the square before the building.

Our readers may now imagine, without difficulty, how in a similar way, snow is formed. If a stratum of air saturated with moisture meets a very cold one, the fog begins to freeze, and becomes specks of snow. They, too, increase while falling, and on arriving upon the earth they are large flakes.

On the occasion of a lecture about the formation of snow in the atmosphere, Professor Dove once told an anecdote, which is as interesting as it is instructive. A musician in St. Petersburg gave a concert in a large hall, where the fashionable world had assembled in great numbers. It was an icy cold night, such as is almost unknown with us; but in the overcrowded hall there was such excessive heat as only Russians can endure. Soon, however, it became too intense even for them. The hall was densely crowded; the throng was alarming; several ladies fainted. An effort was made to open a window, but without success—

the window was frozen fast. A gallant officer devised means; he broke the window in. And what happened? *It commenced to snow in the concert room!* How did this come? The vapor exhaled by the multitude of persons in the hall had collected above, where the air was hottest. The sudden entrance of the icy air through the broken window changed the particles of water into snow. Thus it was this time not heaven, but the upper space of an unventilated concert-hall, that sent down snow.

In a similar way hail is formed in the atmosphere; this we shall consider at more length hereafter. At present we must turn our attention to the influence of these phenomena upon cold and heat; for it is a known fact, that rain and evaporation are not only engendered by cold and heat, but, *vice versâ*, that rain and evaporation, in their turn, engender cold and heat in the air.

# CHAPTER VII.

## HOW HEAT IN THE AIR BECOMES LATENT, AND HOW IT GETS FREE AGAIN.

In the preceding chapter it was shown how warm air produces evaporation, and how cold air causes rain and snow. In this chapter we desire to demonstrate how the reverse may take place, viz., the engendering of cold and heat by evaporation and rain.

Although what we wish to prove in the following is firmly established, yet it is not easy to make it understood. For this reason many educated men, who have read much about "free and latent heat," have mistaken ideas about it.

In order that what we shall explain may be in the reach of every one, we must again choose our examples from life itself, and request our readers to come to our aid with their thoughts.

Every one knows how water is boiled. It is placed over the fire, the heat of which communicates itself to the water and heats it more and more. Now, where does the heat of the fire go? It is taken up by the water; thus to speak, the water absorbs the heat. This explains why a cooking-stove on which a dinner is cooked, does not get near as warm as it would, if the same quantity of fuel had been used without any cooking on the stove. For a portion of the heat being absorbed by the meat, it cannot heat the stove; hence the stove fails to receive the amount of heat that is used in cooking the meat.

What will be the effect of taking boiling water from the

stove and placing it in the room somewhere? Where will the heat of the water go then?

We all know that in this case the water cools down by degrees. The water gives out its heat. Now, it is evident that while on the fire, the water had absorbed heat; and that it gave out that heat on being put in a colder place.

But what will become of the water if it is allowed to continue to absorb heat? What becomes of a pot of water, if, on beginning to boil, it is not taken off the fire? Does such water continue to absorb heat?

Observation shows that this is not the case. Put a thermometer into boiling water; it will immediately rise to 212 degrees; let it remain there ever so long, it will not rise a degree higher. But during that time there was a brisk fire; it is evident, therefore, that heat was continually passing into the water. Where, then, is this heat? It has not remained in the water, or else the thermometer would have continued to rise. It must be, then, that it has passed away with the burning hot steam which has been constantly rising and floating about in the room. Moreover, it is well known that water, when allowed to continue to boil, decreases in quantity. Our housewives call this "*boiling down.*" In truth, however, the water boils *up;* for, if you notice carefully, a part of the water, while boiling, is changed into steam, which may be seen rising from the pot and ascending in the air. The question naturally arises now, where is the heat that the boiling water has been continually absorbing? It has not remained in the water, or the thermometer would have continued to rise. The answer is now evident: the heat has risen with the steam, and with it floats about in the air; or, in other words, the heat has been absorbed by the steam; or, which is the same, the heat has become latent in the steam. Therefore we are correct in saying, *it takes heat to change water into steam.* We know now where the heat has gone; it has become latent in the steam.

The next question might be: Can this latent heat become free again? Certainly it can; and many a good housewife has convinced herself of it very often, though perhaps she did not philosophize about it. When touching unawares the spout of the tea-kettle with her hand she felt as though her hand was wet, and scalded besides. Whence did this come? The hand was wetted by the steam, which, on coming in contact with the hand, changed to water again, but in the same moment, also, the steam gave up its heat to the hand by scalding it. Steam, therefore, when changing into water, gives its latent heat up again; or, the latent heat becomes free.

This phenomenon, which may be witnessed in every kitchen, happens in nature on a larger scale; by what powerful effects it is accompanied, we propose to show in the next chapter.

# CHAPTER VIII.

## LATENT HEAT PRODUCES COLD; FREE HEAT, WARMTH.

He who considers how water when heated is transformed into steam, and how this steam has absorbed the whole portion of heat that was necessary to form it, will easily understand, that places where vapor is formed must become cooler. Just as the fire used for cooking purposes cannot heat the stove, so that portion of the sun's heat which changes the water on the surface of the earth into vapor, cannot heat the earth. Hence it follows, that wherever water evaporates, the air *turns cool,* because the heat, instead of being imparted to the air, is used in forming vapor; this vapor, then, contains the same portion of heat that was necessary to form it; or, scientifically speaking, vapor makes heat latent.

When in summer it is oppressively hot, and a heavy shower comes, it is often more oppressive during the rain than before; but *after* the rain the weather is, as we call it, *cooled off.*

What is the cause of this? After the rain the surface of the earth is wet, and the moisture begins to evaporate. In other words, the rain-water changes again into vapor. To do this, heat is necessary, and is withdrawn from the air and from the surface of the earth; by this means air and earth become *cool.*

It is very agreeable during the summer-time to have the streets of cities sprinkled with water, and it is also very healthy, because the evaporation of the sprinkled water renders heat latent, and thus cools off the air.

The reverse, however, may also take place. As the housewife's hand is scalded when the steam changes on her hand into water, that is, as the steam by turning into water again gives up the heat it possessed, just so acts nature. When vapor in the air turns into rain, it gives up that portion of heat which it had held latent, and hence it it is, that *before* a rain or snow-storm the weather turns warmer.

When in winter it suddenly turns a little warm, that is, when the cold suddenly diminishes, we know that it is going to snow. The only reason why it has become warm is this, that in the air above, vapor has changed into snow, thus giving up its heat, the benefit of which we feel. Thus in summer-time, when the sun becomes fiercest, people say "The sun draws water, it will rain." The truth is, that the vapors in the air change into water, and thus give up their heat; people now think the sun has become hotter.

Another consequence of this phenomenon is the fact, that in countries where there is much water, the air in summer is much cooler, because a great deal of water evaporates there, by which means heat is absorbed or made latent. In winter the air in such countries is warmer, because much vapor is changed into water; thus heat becomes free.

It is evident that all this has a great influence upon the weather—an influence that may be calculated even in advance.

To state an example: The positions of Berlin and London are such, that the summer-heat and the winter-cold ought to be equal in both places. But because England is an island in the ocean, that is, surrounded by large masses of water, the evaporation of water is in London much greater; hence the summer there is cooler. For the same reason rain and fog are much more frequent there, and the winter, consequently, is less severe.

In the course of this work we shall see how similar conditions have very great influence upon whole countries, and therefore often cause, contrary to the rule, cold summers and warm winters.

# CHAPTER IX.

## RULES ABOUT THE WEATHER, AND DISTURBANCES OF THE SAME.

If we cast a glance upon the phenomena of our atmosphere, we find that they are indeed computable, and that the weather in general may be foretold, even for large countries, with some degree of certainty. Nay, there are countries where the weather is not variable at all, but changes at regular periods and according to fixed rules.

In countries near the equator, where the sun's heat is very strong, heat, calm, and dryness prevail during the summer-time. This state of the atmosphere continues uninterruptedly until winter; nor can there be any frost there in winter, because even then the sun's rays fall with but little obliquity upon the surface of the earth. But inasmuch as the sun no longer heats the earth to the same degree, the air ceases to retain the same amount of heat, and as a great deal of cold air is constantly passing in from the poles, the vapor spoken of above is, at that season of the year, changed back into water. Thus, winter there is merely a long, uninterrupted rainy season.

We see that for the warmer countries the rules of temperature are pretty constant and sure; there one is not surprised by irregularities of weather such as occur with us. Summer brings heat, calm, and dryness; winter, east winds, thunder-storms, and continual rain. The rain once ceasing, the sun reappears in a few days, and everything begins to bloom again.

This holds good only for the countries near the equator.

The further you go towards the poles, the more varied become summer and winter, the length of day and night, heat and cold, and consequently, also, the condition of the atmosphere and of the weather proper.

A glance upon the map will convince any one, that it is with us that the weather is most changeable. The reasons for this may now be more closely examined. Our country lies nearly half way between the pole and the equator. From our pole we constantly receive a cold wind, the north wind. And above, in the atmosphere, a warm wind, the south wind, goes continually from the equator to the pole. Through the rotation of the earth around its axis from west to east, the north wind becomes an easterly, that is, a northeast wind; and the south wind in the upper atmosphere becomes a westerly, or southwest wind. The former, coming from cold countries, carries no vapor with it; hence, during northeast wind we have clear sky, or sunshine, but without heat. If this wind occurs in winter, it brings us dry frost; in daytime the sun shines splendidly, at night the stars sparkle brilliantly; yet our breath freezes on our lips. The same wind when prevailing in the first days of spring, causes us, in spite of the glaring sun, to feel considerably cold in the shade.

And it is but natural that it should be so.

The wind comes from the north ; there ice and snow are just melting, and the sun's heat being employed for this "melting business," the air cannot receive much of it.

This kind of weather would be regular with us; but, as we know already, the heated upper air flows from the equator to the north pole; now we live in the very region where this upper air, in its descent towards the poles, at times touches the surface of the earth, thus causing warm currents of air, which occasionally are followed by cold ones.

Near the equator the cold current of air moves below

and the warm one above; while in our regions, both currents meet near the surface of the earth, struggle with each other, seek to repel one another, rush and roll in all directions over the land, and bring us such varieties of weather as will exasperate all weather prophets, and greatly increase the difficulty of scientific solutions in meteorology.

In the next chapter we shall endeavor to prove that this state of affairs, together with the situation of our country, are the main causes of the changeableness of our weather.

# CHAPTER X.

## THE CHANGEABLENESS OF THE WEATHER WITH REGARD TO OUR GEOGRAPHICAL POSITION.

We have endeavored to explain why our weather is so uncertain and incomputable. As we have seen, it has its origin in this, that in our regions the warmer equatorial currents of air no longer move *above* the colder ones, but that they descend here, and pursue their northern course alongside and opposing the colder currents. This often gives rise to a struggle between cold and warm currents. In summer we witness such combats very frequently. The sky is at first bright; the sun sends down his most powerful rays; in the shade we are refreshed by a strong draught, which keeps the sky clear, and free from clouds. Suddenly there comes a calm. Even in the shade the heat now becomes intolerable. The trees stand immovable; no leaflet stirs. The complete calm becomes unendurable, and causes anxiety. "Always a calm before a storm," say the people, and hasten to seek shelter in their houses—and well! for it is not long before a counter wind commences to blow. The weathercock turns round, the dust in the streets is whirled up in eddies, and here and there rises in clouds to the house-tops. Suddenly clouds are seen to form themselves; the trees shake their crowns; the leaves rustle, and before one is aware of it, we have storm, thunder, and violent rain, which cool off the earth.

Whence came this weather; more especially, whence came the calm preceding it, and the whirlwind following?

There were two opposite currents of air, which for a time

avoided each other, but at length met over our heads. Each current at first pressed on the other with equal force, so that they mutually were brought to a stand-still; this we called a *calm*. But such an equilibrium does not last long, for one current must in the end overcome the other; they whirl through one another, raise the dust in high columns, seize the trees and give them a thorough shaking. The cold current changes the vapor of the warm current into clouds, then into rain. The pouring down rain immediately sets free the heat. At this stage electrical phenomena are witnessed, such as lightnings, claps of thunder, and concussions of the air. And this continues until one current of air has carried the victory over the other; not till then does the weather become quiet again.

Besides these opposing currents of air, which come from the north and south, there are other causes disturbing our weather, viz., the geographical position of our country in regard to the east and west.

A glance on the map reminds us that our continent borders, on the east and west, on that immense waste of water, the ocean. We know now that the air above the water is always saturated with vapors, while the air over the land is comparatively dry. And moist air contains heat, dry air does not; both, however, are continually tending towards equilibrium and wish to exchange temperatures from each other. As our dry air is surrounded on both sides by moist air, it is evident that we must more or less partake of both heat and cold; but it moreover accounts for the happy circumstance that we have much rain; hence our soil is well watered, and this is a blessing to any country.

# CHAPTER XI.

## ABOUT THE DIFFICULTY AND POSSIBILITY OF DETERMINING THE WEATHER.

Having now explained the rules referring to the conditions of our weather, and proved that owing to the geographical position of our country, to determine the weather in advance, is difficult, we wish to examine this difficulty a little more closely in pointing out the wrong direction which has hitherto been pursued in the science of meteorology.

The main difficulty in predicting the weather for any given place consists in this, that a change in the atmosphere need not originate in the place where it occurs. Thus, to-morrow's weather in New York is not a consequence of the condition of the air as it exists there to-day; for the air is continually moving, and, owing to many disturbances, is carried over city and country. We have no sure means of ascertaining whence the wind will come to us to-morrow. All we know is, that from all sides currents of air are moving simultaneously; from the north pole a cold current, from the equator a warm one, from the ocean one saturated with moisture. All these winds are in continual commotion, and have the characteristics of the neighborhood from which they come. If from the state of the weather in New York to-day it were desired to predict the weather there for to morrow, one ought to be able to overlook a space of about a thousand miles around; in other words, it must first be ascertained what is the state of the atmosphere within about a thousand miles of the city. Besides, there should be known the direction of all the winds within this wide space, and their speed, and whether they contain much moisture or little. Not without this information could a calculation be made about the velocity with which a change of the weather would take place in New York; what results the meeting of two or more currents of air might call forth; and what kind of weather this might produce there.

Weather, therefore, for the present state of meteorology, is but a subject of investigation into the existing condition of existing phenomena, and not a subject of prediction of coming phenomena. It is true, there are general rules by which a proximate success in predicting may be obtained. If winter begins mild, or, better, if southwest winds and rain prevail till the middle of January, it is very likely that this will be counterbalanced by a northeast wind in the latter part of the winter. The saying, therefore, is correct, "green Christmas and white Easter;" but this rule is by no means infallible, the counteraction may be accelerated by violent storms, or greatly retarded by mild currents of air.

Not before the time that meteorological stations are established throughout the land, and connected by electric telegraphs—a project which to us may seem immense, but to our children will appear very simple and natural—not before that time will a city like New York, for example, receive timely information about the conditions of the currents of air at all the stations. At each of these places the force of the current, its warmth, moisture, and weight will be accurately ascertained by instruments. Then, and then only, we may calculate what currents will meet and where, and what effects the meeting will have. If this be done on Saturday, the Sunday papers will be enabled to state precisely whether the church-goers must provide themselves with umbrellas or parasols.

But not for Sunday alone will this be of importance. It will be long after their establishment, that such weather-stations, connected by telegraphs, will prove their real efficiency and blessing; and our descendants, perhaps, will wonder how we could live without an institution, which to them will appear as simple and natural as do to us gaslights and railroads, which by our forefathers would have been rejected as idle dreams or works of witchcraft.

# CHAPTER XII.

## THE FALSE WEATHER-PROPHETS.

WE wish to speak here a few words about the false methods, that have hitherto been applied to the investigation and foretelling of the weather.

The weather prophecies of the almanac are a disgrace to our advanced age. Those who still print them deserve that their productions should nowhere find sale. We are not of those who expect everything of the magistrates and their orders; but an example should be set to prevent the publishers from dishing up to the people such absurdities.

Some of these wily prophets pretend to read their predictions in the course of the planets. For this purpose, they have divided the planets into two classes, according to their positions in regard to the earth and sun: 1st, those that produce cold, and 2d, those that produce heat. By this means they pretend to prophesy how many degrees of heat or cold there will be every day at sunrise or sunset

When critically analyzed, these prophecies prove to be theoretically and practically nothing but charlatanry.

It is beyond all doubt that the position of the planets is, to state an example, for Boston the same as for the city of Washington; if there are any heat or cold-producing planets, they would have the same effect at Boston that they would at Washington. But this is not the case. Boston has often cold weather when in Washington it is very warm, and *vice versâ.* Besides such a heating or cooling influence of planets would be perceivable on every spot of the earth alike which again is not warranted by facts. On the contrary it often happens that when cold winds are passing over one part of the country, warm winds are passing over another. It is almost certain that cold winters in Europe always accompany warm winters in America; and again, that cold winters in America usually accompany warm ones in Europe. On a closer examination of the facts in the case, we must conclude that, on the whole, weather-prophets take things very easy. Noting the mean heat of each day, and trusting to their good luck, they prophesy one or two degrees above or below. Now, there is no great risk in doing this, and as a matter of course such prophecies are realized one out of two. But at times, almanacs announce an extraordinary increase of cold or heat for a given day, although the situation of the planets does not change suddenly in one day. Then, their predictions very seldom prove to be correct.

In such cases the almanac-makers know how to manage affairs. The country being very large, they send for information to those places where observations on the weather are made. It is almost certain that somewhere in the land their prophesy has come true. Very likely the cold may have increased extraordinarily in the course of a day at New York, Boston, Chicago, Cincinnati, or St. Louis, etc., afterwards the weather-prophets compare their predictions with the results of observation in the various cities, and publish whatever of them are found to have been true.

## CHAPTER XIII.

### HAS THE MOON INFLUENCE UPON THE WEATHER?

THE idea that the moon exercises an influence upon the state of the weather is very general, not only with the people, but also among the better educated. What induces them to entertin it, is not real observation of nature, but a belief which is not without a semblance of truth. If, they say, the moon has enough influence upon our waters to produce tides, it must exercise a still greater influence upon the sea of air surrounding us, and hence it must be of the greatest importance to our weather.

This is, however, an illusion. A long time ago it was proved by Laplace, that tides are caused by the great weight of a liquid. If the ocean were filled with mercury instead of water, the tides would reach a formidable height. Tides, then, do exist in the atmosphere, but in comparison much less than in the water, because the air is so much lighter. It happens that we do not live on the surface of the atmosphere, but in the lowest strata of this airy sea; and in these strata, where the weather manifests itself, the effect of the tides in the upper air is so insignificant, that nothing of it has yet been discovered in spite of most diligent barometer observations.

Learned men have had such a respect for this popular belief, that thorough observations and investigations have been made in order to settle the question.

These investigations were of three kinds:

1st. What influence with regard to heat ana cold has the nearness or remoteness of the moon upon our weather?

2d. What influence has the same upon rain or dryness in the atmosphere? 3d. Has the change of the moon any bearing upon the variability of our weather?

For the reply to these questions, some naturalists have made use of the minutest observations for a period of nearly forty years; during which time the temperature, pressure, and moisture of the air have been measured daily.

These observations have been scrupulously examined; the conclusion arrived at is, that the moon is not quite without influence upon the state of our atmosphere; but this influence is so very small, that it is not brought to bear at all on meteorology.

When the moon is nearest to the earth, it is certainly a little colder than when she is farther off; but the decrease of heat amounts in the average scarcely to one-fifth of a degree, and this is a quantity entirely imperceptible in our weather. As to rain, it is a little less frequent in the time of the moon's greatest distance from the earth; but this difference, too, is imperceptibly small. In one thousand rain-storms there are four hundred and eighty-eight during the moon's greatest distance, five hundred and twelve during her nearest. As to the pressure of the air, it is during the moon's greater distance somewhat greater than when she is nearer, but this difference is still less than the preceding ones, so much so that a common barometer does not even indicate it.

The most thorough investigations have been made about the influence of the waxing and waning moon upon the weather, because it was on this subject that the greatest illusion prevailed. The result here is likewise, that scarcely any difference exists, and that it is a mere superstition for people to maintain, that when the moon changes, the weather changes also. The change in the moon, moreover, does not take place all of a sudden, but with great regularity from day to day, from minute to minute; while the weather, especially with us, changes often very abruptly.

It is therefore certain, that in meteorology one has only to observe the earth and her position with regard to the sun, together with the currents of air and the position of land and water. Other phenomena of the atmosphere may be entirely omitted.

# PART VII.

# OUR ARTICLES OF FOOD.

# CHAPTER I.

## THE RAPID RENEWAL OF THE BLOOD IS AN ADVANTAGE.

Our articles of food are also called *articles of life*, and very properly so; for that which lives in us is, indeed, nothing but food transformed into ourselves.

According to this, it is very easy to determine what a man must eat in order to live; what kind of food can best maintain his health; what constantly renews his working-power; what compensates for the loss he experiences by emission of breath, perspiration, and excretions.

This easy task many have proposed to themselves. They believe they have solved the problem, if they can but prove that all parts of the human body are fed by the blood; and, the constituents of the blood being well known, they believe they have done enough, if they designate that food as the most proper for man which contains the constituent parts of the blood, or which, by digestion, may be changed into blood.

As a general thing this is true; yet it is not sufficient to give the necessary information about the principal articles of our food.

The poor Irishman, who lives almost exclusively on potatoes, has as much blood in his body as the Englishman, whose workmen threaten him with a strike, if they do not earn enough to have a piece of meat and a good glass of beer for breakfast. The Irishman's blood contains quite the same elements that the Englishman's does, and yet their food is very different; and the Irishman is as justly called "poor," as the Englishman is said to be "well fed."

It is evident that the blood alone does not account for this, nor can it do so. There must be other additional items; and these we shall try to learn before we speak of the different articles of food and their worth.

The first principle which we must set up before all others, runs thus: Nutrition does not depend on the blood, but rather on its quick renewal.

The blood resembles the capital which a man possesses. No one can live on his capital without consuming it; he must live on the interest of the capital; he must live by constantly turning the capital over. And so must it be with the blood. The comparison seems so perfect, that we can illustrate our idea best by an example.

Imagine two merchants, each of whom has but a hundred dollars. Both merchants are therefore equally rich in capital. But there is the following difference between them: the one goes to the country twice a week and buys cattle and brings it to market, where he sells it again. By doing this he realizes every time five dollars on his capital. The other establishes a notion-store, buys himself a hundred dollars' worth of goods, which he sells in a month, and thereby gains twenty-five dollars. Now, which of these two fares the better? The notion-dealer, who with his hundred dollars has earned twenty-five dollars, or the cattle-dealer, who gained but five? Most assuredly the cattle-dealer. For while the other has twenty-five dollars to live on, the cattle-dealer has eight times five, or forty dollars. Whence does this come? In a month the notion-dealer turns over his capital but once, while the cattle-dealer turns his eight times.

The same holds good with the Irishman and the Englishman. Both have the same quantity of blood; it is their capital, and the same for both. But the renewal is not the same. The Englishman works vigorously and eats vigorously. When he works, he spends his capital, his blood;

by every blow of the hammer particles of his body are wasted; the activity of his body is great and his appetite is great. He invests his capital again and again in rapid succession, and he takes it in just as rapidly and fares well with it. The poor, unhappy Irishman, however, spends his blood but very slowly; he does not work; he eats potatoes, which, taken alone, are bad food; thus, he invests his capital very slowly and takes it in again very slowly; and though the capital is in both cases the same, its slow renewal is the cause of the Irishman's being miserable, dull, and lazy, while the Englishman is sound in body and soul.

Therefore the blood alone is not all, but its rapid consumption and renewal is the most important object.

# CHAPTER II.

## DIGESTION.

In the preceding article we said that the rapid conversion and waste of the blood is the main point in nutrition In the examination of food, only such articles ought to be pronounced good and healthy, as are capable of *rapidly* replacing the blood lost by work and vital activity. It follows from this, that our chemists do not do enough, when they examine the food and determine its worth merely according to its contents; articles of food must be studied also in reference to the rapidity and ease with which they may be converted into blood.

An article that contains little of what the blood needs, but which converts that little rapidly and easily into blood, is much preferable to food which contains many of the constituent parts of the blood, but turns into blood very slowly and with difficulty.

An example will illustrate this better :

It has been proved chémically, that the husks of grain, the pure bran, contain a remarkably large quantity of vegetable albumen and fat; in this particular, bran is richer than even flour, and a distinguished chemist, Millon, in Paris, in 1849, created quite a sensation by his earnest admonition to use bran no longer only to feed cattle, but to use it mixed with flour, as food for man. He calculated minutely and proved irrefutably, that such food must be considered a great advantage, a real blessing.

Although his investigations and computations were correct, it has since been shown that his proposition is false. In his capacity of a chemist he was right; but the stomach has not as much time and patience as a studious chemist. Notwithstanding bran contains much that the blood can use, yet it is of no avail so long as our digestive apparatus is not organized to perform the change of the bran into blood *rapidly* and *easily*. If bran leaves our body undigested, which happens even to the strongest, then it is certainly more judicious to give it to cattle ; they can digest it well, grow fat and strong upon it, and give us meat, fat, and milk in return.

There is another truth we must constantly keep in view ; it is this : Of two like articles of food, the better and more advantageous one to us is that which is digested, or better, converted into blood, the more easily and quickly.

And there is a third truth, which must not be omitted. Let no one for a moment believe, that a great variety of food is something unimportant and indifferent ; on the contrary, investigations have shown that uniform food is hurtful, while a constant change is very beneficial to nutrition and health.

Nor must we neglect, by way of conclusion, to mention a very important item, viz. : that taste comes in for a large share, and that a judicious assortment and seasoning of the food is an essential part of good nutrition. The husband provides for his wife, it is true; but, on the other hand, the good housewife who prepares healthy, tasteful meals, does in truth perform a great service, and contributes more to the working power of her husband, than most of men are aware.

After these few preliminaries, we will speak now of the articles themselves; in doing so, we shall keep within the limits of practical life, though we run the risk of transgressing here and there into the domain of our good housewives, and of meddling with what, according to their idea, is not our business.

# CHAPTER III.

## COFFEE.

WE come now to consider the various articles of food in detail. We shall take for guide neither the luxurious life of the rich, who, on account of his disordered stomach, constantly tickles his palate with dainties; nor the miserable life of the poor, who, on account of his empty stomach, is bound to find everything palatable. We wish rather to take into consideration the food of that class of people in which the husband works hard to support his family; and where the wife is a good housewife, and cares for the health and strength of her husband and children. In other words, we wish to consider the kind of food called *household fare*, and speak of the meals as taken every day.

It is customary with most to take coffee in the morning.

Now, what are the qualities of coffee? Is coffee an article of food? Or is it a beverage merely to quench the thirst? Is it a means of warming? Or is it a spice? Medicine? Or perhaps poison?

It is strange that science has not yet reached the truth about these questions.

Coffee has been chemically analyzed, and has been found to contain a peculiar element, caffeine, which has an abundance of nitrogen. It is remarkable also that tea has been found to contain an element called theine, which has the same quantity of nitrogen.

As in some countries tea replaces coffee—this is especially the case in Russia, Holland, England, and America—the great and ingenious naturalist Liebig has come to the conclusion that it is nitrogen which constitutes the chief value of tea and coffee as articles of food; and as our blood needs nitrogen, in order to be able to form our muscles and flesh, coffee, according to Liebig, must be counted among the articles of food.

In later times this view has been attacked. Although it is true that nitrogen is very abundant in coffee, and that we need nitrogen to form muscles, yet it can never be the nitrogen which incites us to the enjoyment of coffee. It is the berry of the coffee that contains the nitrogen; a part of it escapes during the process of roasting; a great part is thrown away with the coffee-grounds, so that the quantity of nitrogen actually left in the infusion is exceedingly small. Besides, if we enjoy in coffee only the nitrogen, we pay very high for it.

In the United States, annually about two hundred and fifty millions of pounds of coffee are used; the cost is estimated at twenty-five millions of dollars. Since the coffee itself is not consumed, but only the infusion, it follows that about 100,000 pounds of nitrogen are consumed at a cost of 250 millions of dollars, which is a terrible waste, considering that for this money seven times as much nitrogen could be taken, if, instead of coffee, meat were used, which contains also a large quantity of nitrogen.

The natural sciences, therefore, show among their scholars professed enemies of coffee. They are, from a medical as well as economical point of view, decidedly opposed to its use. Some have even gone so far as to declare it poisonous; a naturalist by name of Zobel proved that it contains Prussic acid, one of the deadliest poisons. Fortunately we know that this Prussic acid is rendered ineffectual by the ammoniac which coffee contains, and which is used as an antidote against Prussic acid.

Be this as it may, we have reason to esteem coffee very

highly. A beverage which has become such a necessity to every nation, is of great importance; and the instinct with which millions and millions of our fellow-men are drawn to its enjoyment, is the best proof that the use of coffee is not hurtful, but advantageous to man; notwithstanding the fact that in some diseases it is forbidden, and that science has not yet succeeded in showing us the real advantage of coffee as a means of food.

# CHAPTER IV.

## COFFEE AS A MEDICINE.

In recent times coffee has been considered, not as an article of food, but partly as a spice and partly as a kind of medicine. Spice it is, inasmuch as it causes, like many other spices, the stomach to secrete an increased quantity of gastric juice. Digestion only takes place when the sides of the stomach secrete a liquid having the quality of digesting food. Owing to this, well-to-do people take after dinner a cup of coffee in order to promote digestion. It is because at night the power of digesting is very much enfeebled—hence the bad sleep after one has eaten something difficult to digest—and because the stomach is relaxed and inactive, that a cup of coffee in the morning refreshes and stimulates the coats of the stomach, and causes there renewed vigor and activity. It is a common observation, that more appetite is felt after coffee than before it. So much for the importance of coffee as a spice. Very justly we ascribe to coffee also a medicinal influence; we consider it a medicine for our mental activity, and for the activity of the nerves.

It is well known that at night coffee dispels fatigue, and that by the use of strong coffee sleep may be banished for a long time. And more; those that are busy mentally, often feel a fresh, invigorating impulse after the enjoyment of coffee; when fatigued with work, they make it a means to recruit their strength. For a similar reason, coffee can animate conversation. When we meet elderly ladies in society, and notice them sitting quietly and talking but in monosyllables, we need not be surprised; they have had no coffee yet! But when, after a little, conversation flows with full force like a rapid stream of water, we may from this safely recognize the mighty influence of coffee; it has loosened not only the tongues, but more—the looks, the hands, nay, the whole body and the whole soul.

Although the mind has rested during the night, we feel in the morning rather sleepy than otherwise, and hence it is, that we are every morning desirous of stimulating our nervous system with a cup of coffee, preparing, as it were, our mind for the day's work. A modern naturalist, as genial as he is learned, Moleschott, ascribes the lately increased consumption of coffee to the greater degree of mental activity, which life in former times did not require to such a high extent as our present age.

We have now sufficiently explained the need of coffee-drinking, and we must confess that all we have said here does not in the least affect our conviction that, according to Liebig, coffee is also nutritive. And no one can help believing this who has seen how old people can subsist on but very little food, provided they can have plenty of coffee. The objection raised, that it would be better for these persons to take the nitrogen contained in coffee in the form of meat, is correct; but, on the other hand, we must stop to ask, whether meat would be good for the stomach at all such times as a cup of coffee is! This would certainly not be the case early in the morning; and if in the coffee we enjoy a beverage which gives us nutriment, strengthens the stomach and at the same time stimulates our mind, we have good reasons to reverence the instinct of man which raised coffee to an essential means of subsistence, and discovered its beneficial influence long before this was done by science.

# CHAPTER V.

## USEFULNESS AND HURTFULNESS OF COFFEE.

SINCE coffee possesses the quality of stimulating the nervous system, it is a matter of course that in many cases its effect is rather injurious. Phlegmatic people, especially, need coffee, and they are fond of drinking it; for a similar reason it is a favorite beverage in the Orient, where its consumption is immense. But to persons of an excitable temperament the enjoyment of coffee is hurtful; they ought only to take it very weak. With lively children it does not agree at all, and it is very wrong to force them to drink it, as is often done; while elderly people, who are in need of a stimulant for the decreasing activity of their nerves, are right in taking as much of it as they choose.

In households of limited means it is often customary to use succory with coffee. We do not pretend to pronounce this, if taken in moderate quantity, hurtful; but we do say, that it is a poor substitute for coffee, and that there is nothing in it to recommend its use. A far better mixture is milk and sugar, and there is good reason for it; both milk and sugar are articles of food. Milk contains the same ingredients as blood, and sugar is changed in the body into fat, which is indispensable to us, especially to the process of breathing. Having taken no food through the night, the loss our blood has suffered during sleep by perspiration, and the fat which has been lost by respiration, must be compensated for in the morning. For this, milk and sugar in coffee are excellent. It is good for children to have a taste for sweetened milk, or milk-coffee, in the morning. We must not find fault with them if they like it. Nature very wisely gave them a liking for sugar; they need it, because their pulse must be quicker, their respiration stronger, in order to facilitate the assimilation of food in their bodies, and also to promote growth. Not that adults need no sugar; but the sugar necessary for them is formed from the starch contained in their food. For this purpose the digestive apparatus must be strongly developed; with children this is not the case; therefore they are given sugar, instead of the starch to make it from. Many diseases, particularly rickets—prevailing mostly among the children of the poor—are the consequence of feeding the child with bread and potatoes; these contain starch it is true, but the digestive apparatus of children being yet too weak to change them into fat, the result is that the flesh falls away, and the bones grow soft and crooked.

But he who, to promote digestion, takes coffee immediately after dinner, does best not to use sugar or milk; for both, so far from helping digestion, are an additional burden to the full stomach, and disturb its labor more than the coffee can facilitate it.

It is very good to take wheat bread for breakfast. Wheat has nearly twice the quantity of sugar and starch that rye contains, and it is besides easier to digest. And as it is our principal duty in the morning to replace as quickly as possible what we have lost during the night, it is a matter of importance to give the stomach such food as is both nutritive and quickly digested.

# CHAPTER VI.

## BREAKFAST.

WORKMEN, even those who must perform hard labor, are sufficiently strengthened by coffee and wheat bread in the morning to begin their work. But to be able to continue it, a more substantial breakfast is necessary, since coffee and bread alone would only replace what was lost during the night. On the continent of Europe it is therefore the custom to take coffee, or milk, and bread very early, and, at about nine or ten o'clock a more substantial meal, a kind of lunch.

Breakfast is with but few the principal meal of the day; for those, however, who rise early it is the one taken with the best appetite. This fact ought to induce all to give attention to this meal; especially those who early in the morning have worked hard already, and those who, mindful of the old saying,

> "Early to bed and early to rise
> Makes a man healthy, wealthy and wise,"

intend not to idle away the precious morning hours.

To him who is in the habit of laboring, and who loves to labor, an early breakfast has a peculiar charm; and, what is yet more important to him, it tastes well. It is customary with us to eat much bread. Bread has as its principal constituents, starch and sugar, and if it has been well baked, a part of the starch is already saccharine, that is, it is nearly transformed into sugar, thus greatly facilitating the process of digestion. French naturalists have lately written excellent treatises about the change which fresh bread undergoes when it becomes old. They prove that bread is most nutritive, and easiest to digest, when about a day old.

Bread is changed in our bodies partly into fat, as all food is which contains starch. But this formation of fat is greatly facilitated, if we take a little ready-made fat with it. For this purpose we eat butter with our bread. Hence we see that some people are wrong when they believe butter to be a mere luxury; on the contrary, butter is a very important article of food, more especially so to children.

The reason of this is, that the fat performs a conspicuous part in the human body; it serves to keep up the process of respiration. The oxygen which is inhaled, decomposes the fat in our body and from it forms water and carbonic acid. The water evaporates through perspiration; the carbonic acid is exhaled again. Now, if there is fat in us, this perspiration and exhalation will diminish it; but this very act of using up the fat preserves our flesh from being consumed in the process of producing carbonic acid and perspiration, which, if there were no fat, would greatly weaken us. Fat, thus to speak, is the spare-money, while flesh is the capital in the body. Fat itself does not make us strong, while flesh does. But where there is no fat, the processes of perspiration and respiration attack our flesh, which, unless abundantly reinforced, begins to disappear rapidly, while our strength begins to decrease more and more.

Thence it comes that lean persons eat much, while we often are astonished to see how little food is taken by fat people. The lean one has no fat to meet the drain produced by perspiration and respiration; he breathes and perspires accordingly at the expense of his flesh, and, therefore, is obliged to continually take in a fresh supply of food. The fat person, meanwhile, does not live on his capital, the flesh and the blood, but on his supply of fat; as it were, he pays expenses from his spare-money, and for this reason loses very little in strength.

From what has preceded, it follows that he who breathes much and perspires much when at work, must eat much fat-producing food, and besides add a little ready-made fat; while he who breathes and perspires little, needs but little of that kind of food. This accounts for the circumstance that in winter, when the air is denser, and therefore one inhales more oxygen and thus uses more fat for exhalation, we must eat more fat food; while in summer every one takes less of it. We know that in cold countries food is taken which, on account of its containing great quantities of fat, would in hot climates produce sickness.

A hearty worker perspires much at his labor, and, in consequence of his increased activity, breathes more than the quiet and sedentary; he must therefore eat with his breakfast some fat—bacon, etc.—because this enables him to prevent his flesh and blood from decreasing. His body will be strong and powerful, and he will at all times be able to earn with his arm more than his stomach costs him.

But let no one believe, therefore, that fat alone is a means of food, and, above all, beware of the mistake that ready-made fat is healthier to eat than fat-producing articles. Fine experiments have been made about the feeding of animals with fat. The results have shown that fat taken alone is injurious, and goes off again without having been of any use to the body; while, on the other hand, fat-producing food greatly assists the fattening of animals.

He who has seen how geese are fattened, will have a correct idea about the process of the formation of fat in the human body. A handful of dough is forced into the mouth and gullet of the goose; during the time of her fattening she is shut up in so close a space that she can neither rise nor walk about. The poor creature is thus deprived of evaporation by perspiration; the process of breathing is rendered very difficult; and, because she breathes and perspires little, her fat does not change into carbonic acid and water, but collects in the body in an unusual manner, until finally the creature is relieved from her pains by being killed. We see that her fat is nothing else than the transformed starch of the dough, which remained in the body without being used. If we should try, however, to feed a goose on pure fat only, she would not fatten at all, but fall sick. Pure fat must only be taken together with fat-producing food. The cause of this is, that only a part of the intestines secretes a juice which can dissolve fat; while the gastric juice in the stomach does not dissolve the fat at all, but allows it to float on the surface, as fat does in water.

Our readers will now find it natural that a workman who perspires and breathes much, should by all means take but little bacon for breakfast; and this he must eat only on those days when he has much work before him; and then he must not eat it without bread.

# CHAPTER VII.

## LIQUOR.

Is it advisable to take a "drink" before breakfast?

This is a question of the greatest importance, and requires a very clear and impartial answer; for which our space is almost too limited.

Liquor is no article of food; if for a moment it were considered as such, we should find that it is even less nutritious than water with sugar in it. What makes liquor a necessary article, especially so to the working-classes, is a certain quality it possesses, a quality just as dangerous as it is good.

Liquor is a favorite beverage because of the alcohol it contains; this is nothing else than sugar which has undergone fermentation. Alcohol may be made from all those plants from which starch can be obtained; for, by the proper process, starch may be changed into gluten, gluten into sugar, and sugar into alcohol. Alcohol therefore conveys more nutriment to the human body than sugar itself, while it has qualities that the sugar does not possess, and which make it an article as popular as it is dangerous. If taken in small quantities, alcohol affects the body like medicine; in large portions, like poison. We are therefore not surprised if partly we cannot do without it, and if, on the other hand, we hear it condemned every day. What makes its enjoyment so very dangerous is, that although it is no article of food, it offers to the hungry a kind of substitute for food, and, what is worse, a substitute which is often the cheapest, and of most rapid effect in regard to quieting one's appetite. It is owing to this that its enjoyment may produce the most fatal and pernicious evils that ever were inflicted upon unhappy man.

Let us now learn the medicinal qualities of liquor, so that we may see that it is natural for it to be a favorite; and by exhibiting the dangers of its enjoyment, we shall succeed best in showing that people are justified in condemning its intemperate use; but it will also be seen that, in spite of the evident hurtfulness, its entire banishment would be a foolishness not resulting in good.

Liquor, if taken in a very small dose, possesses the quality of increasing the quantity of gastric juices. It excites the sides of the stomach, and by this promotes the secretion of the juice by which food is dissolved. It often occurs, that if but a minute quantity of fat has been taken, it envelops the food in the stomach; and as the gastric juice dissolves fat only with great difficulty, this food often remains undigested in the stomach, and nutrition then is carried on but defectively. Digestion, therefore, may be greatly improved, if the stomach is so affected as to secrete a greater quantity of gastric juice; this is often done by means of spice—for example, by putting a little pepper upon bacon or ham. The pepper itself does not help dissolve food, but excites the salivary glands and the stomach, thus increasing the gastric juice which performs digestion.

If fat has been eaten, the same effect may be produced by a little liquor. Indeed, it is even preferable to spice, inasmuch as it contains ether, which alone is able to dissolve fat.

Thus we have seen that liquor is a kind of medicine. And although every one must strive to do without medicine, still he must not condemn it; he should scorn rather the wantonness which throws itself on the mercy of medicine. It is right to oppose the enjoyment of much fat; but if once too much of it has been taken, there is no reason why we should remonstrate against the medical application of a small quantity of liquor. To those who believe that they see in alcohol the evil spirit himself, it may some time or other happen, that even they eat a little too much fat, and then seek relief by taking some patent or other medicine, dropped on sugar. Most medicines used in such cases, however, are nothing but mixtures of sulphuric ether and alcohol; and if alcohol is the evil spirit, he is certainly not changed into an angel by putting him on sugar.

But liquor has yet another effect of great importance.

The alcohol it contains is immediately conveyed to the blood; through this it affects the brain and the nerves, exciting them to increased activity. By also affecting the nerves of the heart, it accelerates the circulation of the blood; this produces throughout the body a more rapid vital activity.

"Wine," the Bible says, "maketh glad the heart of man."

And wine itself is nothing else but an alcohol-combination. The animating element in wine is the same as the one in liquor But it makes man's heart glad; which means as much as, it increases our vital activity; it rouses; it strengthens the weary and him who is exhausted bodily or mentally; it excites the body as well as the mind to more vigorous action. Taken in very small quantity, liquor has the same effect. It is therefore not only good for digestion, but also a prompt remedy for exhaustion. The reanimation, however, produced by the use of stimulants, is by no means a real gain; for he who feels tired and weary is best restored by nature herself. Artificial stimulation is followed by a greater reaction, by which all is lost again that has been gained by artificial animation. Yet many cases occur in human life when there is no time for the natural restoration of strength lost; thus, when it is preferable to complete one's task without delay, without rest until it is finished. In such cases the desire for artificial stimulants is easily explained; then we ought not to condemn a moderate use of them, because that use is necessary.

The wanderer on his travels, the soldier in camp or battle, have often neither time nor opportunity to refresh themselves with a meal, or to recruit strength by a good rest. With them it is important to complete their journey or task, and to rest afterwards. A common workman may, at times, be in the same situation. In such cases a little brandy is of great service. It increases vital activity and courage; in many countries the army is for this reason permitted to use liquor, although, of course, sparingly.

Having now spoken of the medicinal use of liquor, we wish to examine more closely its dangers, and to explain the reason why its enjoyment is to many so great a temptation as often to become a passion.

A slight quantity of liquor taken at breakfast, makes one feel increased vital activity. The pulse beats quicker, the mind is stirred up, digestion easier, and before the food has been transformed into blood, we feel animated to vigorous bodily activity and motion. The enjoyment of spirit fills the long pause between the meal itself and its change into blood. He who feels exhausted and eats, has yet but satisfied the demands of the stomach, without therewith replenishing his blood. It takes a long time, often from five to six hours, before the blood is directly benefited. It is owing to this, that after dinner we do not feel lively, but inactive, disposed to rest. Now, he who after dinner cannot rest, but must continue to work, is anxious to stimulate himself by a dram of liquor, because this will act more quickly than the food he has taken. The spirits he took fill the long pause which exists between his meal and its complete transformation into blood.

Is it any longer surprising, that it is the workmen who

mostly are subject to the use of spirits? No, we are not surprised; we feel sorry that they are not taught better; that instead of imparting to the people a knowledge of things useful to the preservation of health, we constantly remind them of the "devil and hell;" and that in place of teaching them, by the study of nature, how to avoid errors and dangers, we merely try to frighten them with future punishments.

The danger of spirits consists in this, that their good qualities, their advantageous effects, manifest themselves immediately, while their evils appear later. Liquor is not unlike a man whose virtues are laid open to every one; whose vices, however, are hidden, and who therefore is seductive and dangerous. If we wish to warn our fellow-men against such a one, we must not do it by denying or concealing his virtues; on the contrary, we must openly tell all his good qualities; the warning in which we lay bare his vices, will then be more, all the more readily heeded.

True, liquor is a medicine; but, like every other medical remedy, it becomes poisonous in the body of him who puts himself continually in such a condition as to be obliged to use it.

He who wishes to preserve his health, must not try to help nature by artificial means; he will only become weak. To illustrate this by an example: it is a well-known fact, that milk contains all the constituent parts of the blood; but if we were to feed a man merely on milk, those organs given him by nature to digest solid food, would weaken to such a degree that he would fall mortally ill. Man is healthy only when he permits nature the free and unlimited exercise of her functions; if he helps nature too much he may kill himself. It is similar with the use of liquor. The person who only now and then corrects nature, that is, when she actually needs it, is perfectly right. But he is very wrong and harms himself greatly, who wishes to assist nature when she needs no help. Unfortunately, the latter is very often the case, and the prime source of evil. The ignorant, having once had the experience that brandy promotes digestion, thinks it is good for him to continue to help his stomach; but he is greatly mistaken. By accustoming his stomach to secrete gastric juice only after the partaking of brandy, he weakens it; the natural digestion becomes defective through this; and the enjoyment of spirits, at first a medical remedy, rapidly becomes an indispensable necessity, with all its evil consequences.

# CHAPTER VIII.

## INJURIOUSNESS OF DRINKING LIQUOR.

He who accustoms his stomach to secrete gastric juice only after a stimulus effected by spirits, destroys his digestive power. Unhappy man! He is no longer able to digest food, unless he stimulate his stomach with liquor. The already weak stomach is, by this habit, weakened more and more. Soon a small quantity will no longer suffice; a larger portion must effect what formerly was done by the smaller; this goes further and further, until finally the *drinker* becomes—a *drunkard.*

It is well to look at the terrible consequences of such a condition more closely, to obtain a clear idea of it; and to examine all the circumstances which unfortunately produce it, mostly among the poorer and working classes.

The condition of an intoxicated person is to be distinguished from that of a regular drunkard. The former has taken alcohol; it goes into the blood, arrives in the brain, and excites the nerves to increased action. The nerves of the heart are also affected by it, and cause violent beating of the heart and pulse. The blood courses through the veins and rushes to the brain. This produces illusions of the senses, and confusion of sensations; sparks before the eyes; buzzing in the ears; dizziness, which makes the walk unsteady; redness of the skin and eyes; increased perspiration; greater activity in the lungs; a shorter and more rapid breathing; excitement of the mind to anger, and dimness of the faculties of judgment, causing the inebriate to believe that he possesses superior strength. If he begins to move about, these manifestations, and especially the dizziness, increase; the slightest obstacle in the road causes him to stumble or fall; he cannot raise himself to his feet, nor can he sit up; but, lying on the ground, he is unconscious of everything around him; overcome with complete exhaustion—the effect of the reaction—he at last falls asleep; but his sleep does not rest him, although, if sufficiently long, it will restore the unfortunate to consciousness. He now suffers from that peculiar fatigue and lassitude which usually follow intoxication.

To this abject state every one is brought who in the enjoyment of spirits loses self-control. It is an unworthy, disgraceful and disgusting condition; but even the best of men may once fall into it; all the more so, if he is no habitual drinker. Strictly speaking, this subject belongs to another chapter; it belongs to that of intemperance, dissoluteness or bad society. If such a calamity has befallen an otherwise good man, let him amend his bodily ache by a cold bath; and his moral ache by an earnest vow not to do the like again.

Far more serious, however, is the lot of the real drunkard. This belongs to the chapter on nutrition, for it is true, we are sorry to say, that drunkards are produced mostly through want of proper nutriment; and it is always the case that constant intemperance is accompanied by that sickly condition in which the stomach is unable to digest solid food.

In a word, he who has accustomed his stomach to perform digestion only after the use of stimulants, has laid the foundation for drunkenness. With wealthy people, we know it to be frequently the case, that they take something "strong" in order to promote digestion; but the danger is here less great. For if the rich be convinced of his wrong, even at a late period, he can yet proceed in his reform energetically. He can afford to take liquid, easily digestible food instead of solid. He will eat little meat, but that little very savory and prepared in a manner to be easily digested. He will choose but light vegetables. He will flavor his breakfast with caviare and lemon; and at dinner he will relish rich stewed fruit, by means of which appetite and digestion are increased. Should he not feel strengthened immediately after dinner, he has sufficient time to wait till his food is transformed into blood. He takes a nap after dinner, and a pleasant walk in the open air, to get an appetite for his well-selected supper.

Now, all these are excellent means to restore the wealthy man's appetite and digestive powers, even if he has gone so far in drinking as to weaken his stomach. It is not *virtue* and *temperance* that causes the *less* number of drunkards among the rich, but the ready *compensation* they can afford, to cure themselves. And it not unfrequently occurs, that when the rich man loses his fortune, or, in other words, when he becomes poor, he becomes a drunkard. People generally excuse this, saying, "it is from despair;" but the truth is, that now he can no longer afford the costly compensation which previously preserved him from such a fate.

But what will the poor do in such a case more especially the workman?

# CHAPTER IX.

## THE POOR AND THE LIQUOR.

THE poor workman who has accustomed his stomach to perform digestion only through the excitement of a previous stimulant, cannot, even if he knows the miserable condition he is in, abandon this bad habit without almost superhuman efforts.

Working makes him hungry; but his stomach not being able to digest solid food, eating becomes disagreeable to him. His relaxing strength, however, demands support. His vital activity is suppressed; he must have a fresh supply of strength to be able to work and earn his living. To accomplish this, he knows no other means than liquor again! For, unfortunately, experience has taught him that spirits not only stimulate him for the moment and increase his vital activity, but that they can also be to him a kind of substitute for food.

It was not until quite recently that science told us how and in what manner the use of spirits may actually promote the working power of the starving. It is of the utmost importance to obtain a correct idea of this.

Work promotes evaporation and respiration. Evaporation, however, that is perspiration proper, is nothing but a part of the food we have taken, and which is thus secreted from the body. Precisely the same holds good with the breath we exhale; it consists of carbonic acid, which is likewise tormed from the food we have taken. A man in state of rest does not perspire and breathe so much as the man at work; therefore he needs less food. If, on the other hand, a person works without taking food, the perspiration and carbonic acid of the breath are formed from the muscles of his body; for which reason he must greatly decrease, both in strength and volume. We must bear in mind, however, that it is one of the qualities of spirits to be decomposed in the body very easily into water and carbonic acid; the water is then secreted in the form of perspiration; the carbonic acid, by exhalation. Thus, if a man works without food, he becomes reduced immediately, because perspiration and breath are supplied from the flesh of his body; while if he drinks liquor, perspiration and breath are formed from the liquor itself, instead of his body, which thus, partly at least, remains intact.

This is the solution of the great problem, viz., "How can drunkards live a long time on nothing but spirits, and, moreover, how can they work?" We know it now; liquor furnishes them the material for perspiration and breath; and their body is not nearly so much taxed as would be the case, if they were to take no spirits at all. Since, then, the drunkard cannot eat, and even if he could, would not be nourished, because food passes through him undigested, he must needs continue taking spirits even if he works but little. Spirits help him at his work, and save his body from being consumed.

That spirits are no articles of food, has been known long; but it was not known until recently, why spirits can be a substitute for food, or, more correctly, a kind of *saving of food.*

Unfortunately, liquor is as deplorable as a substitute as it is fatal as a means of saving. It is only calculated to entirely destroy the doomed man that uses it.

Now, is it not more judicious to understand the reason why the drunkard cannot abstain from spirits, than to endeavor to reform him merely by "prayer" and stories about the "devil in the alcohol?" And is it not of the highest importance to all, that the friends of humanity should take care that the workman has good and healthy food, and that he be always able to earn enough, so as not to be obliged to replace bad food by liquor?

The workman who has nothing but potatoes to eat, is bound to become a drunkard. This food is insufficient to afford him a proper quantity of carbonic acid for the purpose of breathing; he therefore must draw for this from his body, and, since he must needs work for his living, he takes to spirits to save his body from being consumed. Many an "Apostle of Temperance" would, in a similar situation, act no better. For this reason let us all provide healthy food for the working class; intemperance will then greatly diminish.

Owing to the importance of the subject we have spent much time over "Breakfast," and the chapter on "Spirits" connected with the same; but we could not help it; nay, we must ask our readers' pardon for continuing the subject. We propose to touch upon the sad consequences of intemperance, and desire to give the wives of the workmen a hint, by which they may succeed in checking the vice of their husbands and the misfortune of their families.

# CHAPTER X.

## THE CONSEQUENCES OF INTEMPERANCE AND ITS PREVENTION.

THE digestion of the drunkard, as we have seen, is greatly impaired; the process of nutrition entirely changed. There is a change in the tissues of the interior of the body. The inner organs are encumbered by fat; even under the very skin, layers of fat are formed. It is this that gives the drunkard that bloated appearance, which is very characteristic, and an evidence of the fact that the evil has reached a high stage. The stomach and the heart, the latter now much enlarged, are in an unnatural manner enveloped by fat. The action of the heart, at times immoderately increased, at times fearfully lessened, causes the blood to rush impetuously even to the finest blood-vessels of the skin, and to widen them considerably. Hence the reddened face of the drunkard. The chest being overburdened with fat, the lungs are unable to expand properly, and cannot therefore feed the blood with a sufficient quantity of oxygen, which would make the blood red; therefore we notice that the drunkard's blood is of a bluish color; his nose is blue, his lips, and often his whole face, have a bluish hue. His mind is always clouded, the activity of his nerves partly increased, partly weakened; his hands begin to tremble, and become unsteady; soon his very feet refuse to serve. His breath is in the beginning saturated with alcohol, so that it can be smelled; in a little while perspiration, nay the whole body, is imbued with alcohol, and cases have been known in which the body, on coming in contact with fire, began to burn, as a wick dipped in alcohol, inflicting a terrible death upon the unfortunate victim. Many die from apoplexy or paralysis of the brain, in most cases preceded by delirium tremens. When it is considered that all this has its beginning only in this, that the unhappy man has accustomed himself to promote digestion by means of spirits—when this is well considered, no one will find it strange that we wish to discourage from the use of liquor everybody, especially, however, those among the laboring classes who work with fire. He who takes proper care of himself will always know how much of spirits he can take and when he must use it; then, and only then, the enjoyment of the article in question cannot be considered a crime.

It is difficult to present to our readers a general rule for temperance, yet we may here state a *principle*, the earnest observance of which we heartily recommend.

There are many people who say: "I can stand a little liquor very well." They mean by this that a little liquor does not intoxicate them. But this is a dangerous standard to take. Not the possibility of *intoxication*, but the welfare of one's *stomach* should be consulted. As long as breakfast can be digested without the use of spirits there is no danger, even if after having eaten fat, bacon, etc., a desire for liquor should be felt; but when a person must needs take spirits after his breakfast in order to be able to digest it, then the danger becomes imminent, and it is high time to consult a physician about this seemingly insignificant circumstance; it is best to tell him frankly the object of the visit, viz., the desire to avoid the cheap remedy, the liquor. If the physician be the right man he will gladly spend advice and help.

In such cases, however, the housewife can do even more than the doctor.

The attentive housewife will notice the bad condition

of her husband's stomach, and if she is judicious and wishes to be the benefactress of her household, she can, by a small sacrifice, easily prevent great misfortune. Above all, she must bear in mind that only a well-fed husband can support her and her children. It is a shame that we often see a housewife treat her husband in this respect worse than a horse. The owner of a horse knows that his horse cannot render him good service unless he feeds the animal well; why should woman not comprehend that man, her husband and provider, must be properly cared for? Let every good wife bear in mind, that if her husband takes to drinking, it is mostly owing to her own bad and careless management of her kitchen; let her hasten to remedy the evil. Although it may cost her a sacrifice, yet she owes it to herself and her family to provide her husband with a cup of broth, well seasoned with salt and pepper, when his stomach is weakened. At times she may surprise him with a favorite dish for breakfast, which he will eat with a relish. And let her be especially careful not to cause him grief or anger at his return home, but let her rather prepare for him a good savory dinner, for which he then will save all his appetite.

Such and similar insignificant acts of womanly kindness preserve often husband, wife, and children from disgrace; while the dutiful wife earns the esteem and gratitude of her family and of her country. This is a merit which in course of time will be duly rewarded.

## CHAPTER XI.

### DINNER.

We wish to speak now of dinner, the principal meal of the day. Here, too, we shall take for standard neither the unhappy poor, who must eat what little he can obtain; nor the opulent rich, who finds a pleasure in eating what others cannot obtain. We shall take for base the plain household of the citizen, who takes healthy meals in order to strengthen him for renewed activity.

What may have been the reason for putting the principal meal in the middle of the day?

It was done for the reason that eating, too, is a labor; a labor which requires rest. Now bodily fatigue and appetite constantly keep pace with each other; they manifest themselves in the body in intervals of three or four hours. Since, then, we must rest at noon from the fatigue of the morning's labor, it is best for us to use this time of rest for our dinner; all the more so as the labor of eating ought not to be performed during manual labor. And because just at the middle of the day we rest from our labor and prepare ourselves for the afternoon work, it is natural that we should eat our principal meal at that time.

But this meal needs to be prepared carefully. The housewife is chained to the kitchen, because this meal is distinguished from others principally in this, that it is usually taken warm.

The question arises in the first place, Why must food be cooked? Is it not more natural to take the food as nature gives it to us? Why does man eat nothing raw except fruit? Why does he take such pains to grind, bake, boil, fry, etc., while the animal can live without all this? Again, whence does it come, that man is so very dainty in regard to eating and drinking, and that he uses an infinite variety of articles of food, as does no other creature in the world? Are there not animals that live on meat only, and others that live only on plants? Why, then, does man need mixed food, that is, partly meat and partly vegetable food?

To all these questions there is but one answer.

Nature herself has pointed this out to man; and experience, the natural instructor of mankind, has taught man how he can do best what nature wishes him to do.

The human stomach is so constituted that it can digest but very little of raw food. Just as the nutritive part of the pea is enclosed by a *hull*, so in every organic food the nutritive element proper is contained in a hull, called *cell*. The nutritive element of the potato, for example—the starch—is enclosed in millions of small cells, which are indigestible for our stomach. By means of good magnifying glasses, these cells, invisible to the naked eye, may be plainly seen. If the potato were eaten raw, these cells, together with the nutritive element in them, would leave the body unchanged. But if the potato is boiled, fried, or baked, the cells, by their expansion from the heat, burst, and thus allow the starch to be free. Now, while animals have been given a digestive apparatus strong enough to dissolve the hardest cells—pigeons, for example, swallow and are able to digest raw pease—man has been endowed with intelligence which enables him to prepare his food artificially.

Cooking, therefore, is as natural to man as the act of chewing; for chewing, the crushing of food with the teeth, on the part of animals that live on plants, is nothing but the tearing asunder of cells. Animals that have no teeth, birds for example, possess immensely strong powers of digestion. It would be as unnatural for the ox, who has good teeth to crush peas with, to swallow them entire as the pigeon does, as it were unnatural for man to take pease raw while he has the means of cooking them.

We often call *art* what really is *nature* in man; for his mental gifts are natural to him; women, therefore, when they perform the art of cooking, practise a natural art.

## CHAPTER XII.

### NECESSITY FOR VARIETY IN FOOD.

LET no one believe that it is from mere daintiness that man is fastidious in regard to food, and that he lives on a great variety of victuals.

The human body is the transformed food which he has eaten. It is quite correct that man can live on bread and water a long time; but man's nature is so varied, his qualities are of such numerous kinds; his character, his impulses and passions, his wishes and desires, his thoughts and labors, are so infinitely varified and so much exposed to change, that man's body, the bearer of all these elements, must also be formed from material of the most diversified kind.

It is a common observation that animals which take uniform food are very much poorer in mind than those animals that feed upon richer and more various kinds of food. Nay, it has even been proved that the character, the whole nature of an animal may be completely changed by its food. Very properly, therefore, does the genial naturalist, Moleschott, begin his excellent treatise, "Our Articles of Food," with the following words: "Food has made the wild-cat our house-cat;" thus showing that food may completely change the character of an animal, and more, it may even change the animal's body. And if civilized man is a being of a higher order, more spiritual and more intellectual than the savage, we can ascribe it to no other cause than the impulse his food gives him, not to sink down to the savage, but, by varying his food as much as possible, to bestow upon his body many superior qualities.

Nature herself has undeniably impressed upon man, that she wishes him to take nourishment of different kinds.

Those animals that live upon plants, and such as feed solely on meat, are entirely different from each other in regard to their bodies. The teeth of the former, the herbivorous, are broad and flat on the top, like our molar teeth. They serve to crush vegetable fibres and to chew the cells which contain the nutritive element; while the other class, the carnivorous, have but pointed teeth, like our eye-teeth, to tear their food asunder. The stomach of the herbivorous is also different; it comprises several divisions which have various functions. For blood is not so readily obtained from vegetable as from animal food, which itself contains ready-made blood. Herbivorous animals are for the greater part ruminators, that is, their food passes from the first division of the stomach back into the mouth, where it is masticated a second time; this is called "ruminating." With the carnivorous this is not the case. Finally, the intestines of the herbivorous are long, because there the final change of the food into blood takes place; a process requiring more time with vegetable food than with animal. For the same reason the intestines of the carnivorous are short, the blood to be formed being already present there.

Considering the fact that man has sharp teeth in front, at both sides pointed teeth, and in the rear of them molars; that his stomach is adapted to the digestion of both vegetable and animal food, and that his intestine is so constituted as to be able to digest and change into blood both kinds, we can no longer entertain any doubt that nature herself bids him to change his food constantly, and to take in such as is of the most varied kind. If, in addition to that, we recollect that exclusive animal food renders an animal wild, quick, and sly, while vegetable food makes it tame, enduring, and slow in mind, it will not be denied that food exercises great influence upon the nature of a being, and it will now be readily understood that it would be a sin, if man were to be forced to take uniform nourishment.

The example of the cat is very instructive; it teaches us that change of food has transformed her into another being, mentally as well as bodily. The wild-cat has short intestines and is an animal of prey; the tame cat has long intestines, and betrays her old character only now and then by cunning and slyness. We also learn from this, that variety of food produces variety of bodily and mental qualities; and lastly, it may be inferred that nature, having fitted man for this variety and given him such diversity of mental capacities, wishes also that his food be well selected and of the greatest variety.

These short remarks enable us to pass to the principal dishes themselves; first to those constituting the principal meal of the day, the dinner, for which very justly the greatest variety of food is chosen.

# CHAPTER XIII.

## BROTH.

Soup, meat, and vegetables are the principal dishes of a plain household dinner.

When examining this more closely, we find the selection so judicious that we may well admire the tact of woman, who discovered it long before science did.

The good tact of woman does even more yet; it selects the dishes in such a manner that they mutually compensate for their wants, that is, that each offers to the body what is wanting in the others.

The principal dishes composing a meal are divided into fat-producing and flesh-producing ones. All farinaceous diet provides the body with fat; all albumen substances, with flesh. To support the body, however, it is also necessary to give it salt, from which bones, hair, nails and teeth may be formed.

Our domestic wives, indeed, look to all that. Long before scientific men had investigated the necessity for nutriment of the kind, all-providing woman had arranged culinary matters so as to be able to satisfy all the demands of nature. But not only the proper selection of articles of food,—the way and manner also in which they are cooked and served, are of prime importance to a proper nutrition; and we maintain that household fare may justly be regarded as a guide for scientific investigations.

A judicious housewife will first of all place meat on the fire, to have good soup and well-cooked meat. She will prefer beef to any other kind, because it contains but little fat and much albumen and animal fibre; for this reason it makes better broth, and still preserves strength enough to be a healthy, strength-giving dish.

Besides, meat, by cooking, becomes more nutritive, inasmuch as its digestibility is greatly facilitated. One of the most important tasks of the cook consists in promoting one's digestion; in other words, in saving the stomach labor. Flesh in its raw state keeps its nutritive elements shut up in cells which are gluey. By boiling it, the gelatine becomes soft and mixes with the water; hence it comes that broth is glutinous, and, if allowed to cool, becomes thick and like jelly. This substance is in part very nourishing; it is often obtained from bones and cartilages, and then sold under the name of "bouillon-tables," which, when boiled in water, make a tolerably good soup. Thus we see that the first object of all cooking is the dissolving of the cellular tissues. Not before this is done do we obtain the real nutritive element of the flesh, which then is taken up by the stomach all the easier, inasmuch as it has thus been well prepared to be easily changed into blood.

But before the meat reaches the boiling-point, albumen is separated from its surface and mixes with the water; it is this which gives broth its real strength and nutritive power. Afterwards, when the water boils, this albumen condenses; the broth becomes white, as if containing the white of eggs; from the inside of the meat flows continually more and more albumen into the broth, and makes it stronger and stronger. During this time, moreover, the fat parts of the meat melt, and its salts are also dissolved in the broth; hence a great deal of the most nutritive parts of the meat goes over into the broth; and although much of the strength of the meat has been withdrawn, still there is much of it left yet, and the meat has now become easier to masticate and easier to be digested. We need not add that a sufficient quantity of salt is thrown into the soup, which quickly dissolves in the water; but in the same degree that the meat excretes a part of its ingredients and gives them to the water, in the same measure does the meat absorb salt. By this it becomes not only more tasteful and digestible, but also more nutritive. It was not until recently that the importance of salt as a nutritive was recognized; this cannot be otherwise, for the tissues of the human body, as well as its blood and cartilages, need salt for their formation and support. Who does not know that every farmer gives his cattle salt from time to time, so as to improve their strength and general health?

Our readers will readily understand now, that the weaker the broth the stronger must be the meat, and *vice versâ*. It often occurs that we care less to have good broth than good beef. In such cases we must not put the meat into cold water, but into boiling water. So soon as the meat is thrown into boiling water, the albumen on the outside coagulates, surrounding the whole piece as it were with a hard crust, which does not permit the nutritive parts of the inside to escape. The same effect is produced by the roasting of the meat in an oven, although here it is not covered by water. It is more judicious, however, and more important for the household, to make good broth, and to let dinner commence with it.

For he who has been at work all the forenoon, needs such food at first as will not cause his stomach too much labor; and soup is that food. Let every good housewife bear this in mind.

# CHAPTER XIV.

## WHAT IS BEST TO BE PUT INTO SOUP?

The answer to this question will be "Something farinaceous," and, indeed, no better answer could be given.

Broth contains gluten and albumen, both of which are changed in the body into flesh. Not only the animal part of our body, but chiefly the active, working part of it requires nutriment that can be transformed partly into fat. Breath and perspiration, so unavoidable in labor, are supported by means of fat in our body. This explains why fat people perspire more than others; why fat people get out of breath sooner than lean persons; why the other sex, who are more apt to become fat than men, perspire more; and why children, because they run about much, and hence need more breath and perspiration, usually prefer bread to meat.

As has been said, broth, which contains only such ingredients as are intended to produce muscle-fibres, may well be mixed with something farinaceous, which should be thrown in and boiled with the soup, in order to promote the formation of fat in the body. It matters little what may be chosen for the purpose—flour, groats, barley, rice, or potato, or any other article; provided always it contains starch; for this becomes saccharine even when boiling; it changes in the body into acid of milk, and lastly into fat. Perhaps it is advisable to use that which contains most starch. Rice, for example, has much of it; probably this accounts for the fact that lively children are very fond of it. A hundred pounds of rice include eighty-five of starch; while a hundred pounds of wheat contain but about seventy-four pounds. A judicious housekeeper will know very well that a less quantity is taken of rice than of flour. The various kinds of farina and barley possess but about one-half the starch of rice; and potatoes are so poor in that, that five pounds of potatoes yield no more starch than one pound of rice. All this is a matter of great importance to our housewives.

The usefulness of soup-material lies, however, not always in its great nutritive capacity, but very often in the facility with which it may be cooked. Thus we cannot boil rice in the broth itself; it must, to loosen its cells properly, be boiled first in water; this takes a little over half an hour, and requires of course a place on the fire, and hence more fuel. The cell of the farina or pearl-barley, on the other hand, was crushed already by the grinding; therefore it needs but little attention, and may be boiled in the broth itself without any loss of time. When making scientific observations on food, such circumstances must not be overlooked; for time and fuel cost money, and may, in the eyes of practical housewives, raise the price of the article too much; while to a scientific man the same article may appear very cheap.

There are other viands which, though not very nutritive, are yet very popular and in common use. As an example of this class, we may give the potato.

That the latter is poor in starch, was stated above. Its extensive use is surprising, when we consider, that, according to calculation, the little nutriment obtained from the potato is paid more highly for than that of flour. And yet there is good reason for the extensive use of the article. Its preparation, in the first place, is an easy one, especially when the potato is boiled whole, without being peeled. This is a great convenience for the housewife, who, besides the time devoted to the house, needs time for work from the proceeds of which she may support herself. She values, therefore, any dish which can be prepared with little expense of time and money; more than any other article may the potato be said to possess this quality. From it a meal can be prepared in half an hour, and the cook need not watch it constantly; potatoes do not boil over. Besides all this, there is another advantage, and it is this which makes it a favorite even with the rich; already, when boiling, its starch is transformed into sugar, giving the potato a more pleasant flavor than any other cheap dish can be said to have. How easily the potato starch is converted into sugar may be noticed best in half-frozen potatoes, because there the cells containing the starch burst during the process of freezing.

# CHAPTER XV.

## LEGUMINOUS VEGETABLES.

The greens which we put in soup cannot be considered nutriment, but rather a kind of spice, and perhaps also as a means of giving us the benefit of some medicinal qualities which they in part contain. We will dwell no longer on this subject, but proceed to the most nutritive articles of food we use, viz., the leguminous vegetables.

Pease, beans, and lentils are so extremely rich in fat and muscle-forming elements, that in this regard they excel bread and are almost on a level with meat. No wonder, therefore, that they are very favorite articles if well cooked, when we consider the fact that they are so very cheap. Where people are too poor to buy meat every day, legumes must not be found wanting. They play a great part in barracks and prisons; and in order to keep pace with the immense progress gastronomical science has made, one of the above-named articles ought to be used in those establishments on all days on which there is no meat.

The element common to all three is called legumine. It is richer in starch than bread and contains nearly three times more of it than the potato. Partly legumes contain also ready-made sugar; this may be tasted in green pease. Besides this, their flesh-forming parts are in greater quantity than those of other plants, while their quantity of water is less, and it is therefore not advisable to take them dry. New pease and beans have, moreover, the advantage of being eatable together with their hulls and pods, as these, when yet green, contain likewise sugar and starch.

But we must recommend, above all, not to eat the hulls of dried legumes. This may be avoided if, when boiled, the cook crushes them and strains them through a coarse sieve, by which process the hulls are left. If this is not done, we run the danger of disturbing the functions of the body, inasmuch as these dry hulls are dissolved neither by the saliva of the mouth nor the gastric juice of the stomach.

Most every one that once in his life had culinary labor to perform, is acquainted with th fact that the cooking of legumes is often accompanied by a peculiar circumstance. Pease sometimes may boil by the hour without getting soft; it happens even that young pease, soft by nature, become harder and harder by boiling; while, at other times, the same pease have become soft and burst open after but half an hour's cooking. The reason of this lies not in the pease, but in the water they are boiled in. Our housewives undoubtedly know, from the experience of their wash-days, that there is hard water and soft. Soap, when put in hard water, breaks into small pieces, while it dissolves in soft water completely and forms a slimy liquid. Science has solved this mystery: spring-water contains lime, which combines chemically with the fat in soap and forms with it an insoluble element; while rain-water contains little or no lime, and therefore dissolves soap. The same is the case in regard to the legumine. The lime in spring-water, which settles on the bottom of vessels as sediment, combines with some constituent parts of the pea and forms a very hard, indigestible body; rain-water, however, dissolves legumine completely.

It must now appear evident to all, that much fuel and nutritive element is gained by cooking pease, beans, and lentils in soft water. To comfort those who, on the plea of uncleanliness, are opposed to rain or cistern water, we desire to state that rain-water when poured through linen or cotton cloth is not in the least impure; especially if it be allowed to stand quietly for a few hours and then have the scum removed from its surface.

Pease, beans, and lentils produce in the healthy body blood, flesh, milk, and fat. By their being strained through a coarse sieve they lose such disagreeable qualities as, for example, the bloating they produce in the body, which makes them very unpopular with many.

Another great advantage in leguminous vegetables lies in this, that they contain phosphorus, a principle needed for the formation and preservation of the bones and brain; therefore we may justly maintain that legumine is good for the body and mind both.

# CHAPTER XVI.

## MEAT AND VEGETABLES.

It is an old German habit to consider meat and vegetables as belonging together.

In the common kinds of vegetables there is very little nutriment. Nearly nine-tenths of the weight of cabbages and other varieties consist of water. There is therefore but little left for nutriment proper, as, for example, vegetable albumen, gluten, vegetable fat, starch, and sugar. It is only such vegetables as turnips, etc., that contain much sugar, for which reason they are well adapted for children and convalescents. In fine, if nutriment alone were considered, the enjoyment of our common vegetables would be nothing but a luxury.

In truth, however, they possess elements which make them very beneficial to man, if he takes them together with meat. They contain organic acids—like fruit, which for this reason is so universally liked—and have the quality of preserving in a state of dissolution the soluble albumen of the meat. Thus they save much labor to the digestive organs, and accelerate the transition of meat into chyle. Hence the well-known fact, that after dinner, though we can eat nothing more, yet we like to taste some good raw fruit, or cooked fruit of any kind. Vegetables are taken for a similar purpose, and are therefore very healthy when eaten with meat.

But why is it that our housewives often serve vegetables *before* they do meat, and fruit *after* the meat?

Very likely they themselves do not know why, as is the

case so often; yet they act here, as in many other things, with wise instinct. Fruit contains organic acid, which, in a ready-made condition, is very beneficial to us; it needs only to be taken up by the stomach. We do well, therefore, if we take fruit after the meat, and allow digestion to go on with it. From vegetables, however, this acid is only produced in the stomach, and during the process of digestion. If taken before meat, the acid may promote the digestion of the meat; while if it is taken after the meat, the acid comes much too late to be of any benefit. This explains the fact, that vegetables in which this acid has been produced by fermentation—as is the case, for example, with sour-crout—are usually taken together with meat.

Another great advantage of vegetables is, that they are rich in mineral salts necessary for the health of the body. There are ingredients in the various kinds of vegetables, of which it may scarcely be believed that they can be eaten, for they belong to the metals and metal combinations; as, for example, chlorine, iron, potassium, and natron; these play an important part in the body. It is, therefore, not surprising to us that a judicious physician will more often prescribe a good vegetable than medicine; and one ought to be thankful to him if he sends people more to the market than to the drug-store. There are, indeed, many diseases successfully cured by such organical remedies, which only nature knows how to prepare. To mention but one remedy, spinage, so highly beneficial to children and young girls of very pale appearance. Their green-sickness takes origin from a want of iron in the blood. Though every physician is able to prescribe medicine which contains iron, yet the effect of such artificial inorganic remedies is often very doubtful; while spinage itself contains iron, and therefore offers a better organic remedy, and food.

Meat and vegetables are sufficient for the body. There

is not need of much meat. From six to eight ounces a day constitutes the quantity sufficient for a man. Meat and vegetables compensate each other's wants; the former is poor in water, the latter rich; vegetables are wanting in albumen, which is found abundantly in meat. This happy circumstance is favorable to the formation of that mixture of elements essential to the preservation of the body.

Household fare, according to what we have seen, is precisely what it ought to be, and does not, as some people are inclined to think, result solely from the whims of the housewives. Thus is proved again what we have said above, viz., that the natural instinct and tact of woman have, by long years of practice, been guided by a better and more practical course than science itself.

There are some other important articles of food, but we must keep them for "Supper;" and our readers will no doubt be very glad if we conclude this chapter, and treat in the next one the question,

"Is it good to take a little nap after dinner?"

## CHAPTER XVII.

### THE NAP AFTER DINNER.

An old adage says, "After dinner thou shalt either rest or walk a thousand steps." Habit, however, has modified this very much; for people nowadays neither rest nor walk; but, if they can, they lie down and slumber. Now, it is true that sleep does not belong to the articles of food. We might despatch the question of the nap after dinner here at once; yet, if it has any influence upon the digestion of food, it is of enough importance to merit a few words.

It was mentioned before, that eating and digestion are a labor. To many it may be the most pleasant labor, to others even the only labor of their lives; but be this as it may, it is certainly a labor for all and every one; and it is important that during the process quiet should be enjoyed. He who thinks he gains by not taking enough time for eating, or he who takes his dinner while working or moving about, loses actually more than he even thinks of winning. The activity without disturbs seriously the activity within. The perspiration on the surface of the body withdraws moisture from the inside of the body to such an extent as to diminish even the saliva in the mouth, so necessary to digestion. Have not all of you had the experience, that when fatigued you feel dryness in the mouth; that you feel as if a piece of dry bread would not pass down, but remain in your throat? And as with the saliva, it is with the other digestive fluids; if there is any want of them, the food we have taken lies in the stomach like stone.

It is therefore desirable to take a short rest before dinner, not to perform any kind of labor whatever during the same, and, above all, not to exercise the body immediately after dinner. Eating is an inward work, and should not be accompanied by any labor without. As an additional proof of what we said above, it may be stated that, as probably many of our readers know already, even in the hottest summer, perspiration diminishes after dinner. This will convince all, that when the digestive apparatus is at work, the outer organs ought to be at rest. Once more, then: before and after dinner we need rest, and it is this rest which renders us indisposed to labor and makes us feel sleepy.

On the other hand, we must take but a short slumber. Those who have accustomed themselves to sleep after dinner, feel that half an hour's slumber is all that is needed, and that they even feel weary if they have slept longer.

The reason of this is, the process of digestion is properly carried on chemically by the food, being dissolved through the gastric juice. This digestion, however, is greatly promoted by the motions of the stomach, which tosses the food about from one side to the other, mixing it entirely, and finally making a large ball of it, whose various ingredients are, as it were, fused together. This process needs rest on our part; during it sleep is sweet and agreeable. But for the further digestion of food, energy is needed, which we have not during that sleep; therefore its want makes our prolonged sleep uneasy, or renders our digestion imperfect. This latter may be felt by every one who goes to bed with a full stomach. His sleep during the first hour is undisturbed and pleasant, because it is favorable to the first stage of digestion. But after that, sleep is very uneasy; weariness and complaints about bad digestion follow, and the imprudent person rises next morning with headache, coated tongue, and indigestion in the stomach.

From what has preceded we may conclude, that a short nap after dinner is conducive to good health; while if taken too long, it will produce the contrary effect. Dizziness in the head and fetid taste in the mouth are sure signs of one's having overslept one's self, and he who has been so imprudent must animate his system—not by liquor, but with a glass of fresh water; or he must, if he feels very heavy, wash with very cold water. For this is the moment when digestion needs activity more than anything else; the above symptoms are the indications, and man should consider them as the summons of nature, who calls to him, "Thou hast eaten and reposed; go, then, to thy labor; this is the time!"

Let every one obey her call, and there will be less sickness.

## CHAPTER XVIII.

### WATER AND BEER.

During the forenoon a general desire for food is felt, while in the afternoon thirst is more common, in which case the best and most natural beverage should always be water.

Properly speaking, water is no article of food, if by that term we understand only animal and vegetable matter. Water is no organic, but a mere chemical agent. But if man were to consume no water he would perish. Therefore water is essentially necessary to man, although it does not satisfy his appetite; for it serves to liquify our food in the body, and our blood must contain a greater quantity of water than is furnished us by food, although this itself contains much water.

Without water there can be neither digestion nor nutrition, nor formation of blood, nor secretion. Furthermore, it is remarkable that the most active of the human organs, the brain and muscles, contain the most water; we are therefore obliged, although we are aware of its containing no nutritious elements, to call it a nutritive; all the more, since it is well known that we can be longer without food than without water.

This element plays a great part in the body; it is used in three ways. In the first place, the ingredients of water, hydrogen and oxygen, combine with the food, and effect its digestion. The starch which we eat in farinaceous and vegetable food cannot without water be converted into sugar. And the latter being transformed into fat, we should have no fat if we took no water, though it may seem strange that water should make us fat.

And there is the second task, viz., the preservation of all the fluids necessary to our body. This, also, is performed by water; and as they are excreted their loss is compensated for by water. We lose it constantly by breathing, perspiring, and urinating; therefore we must continually take it anew. Those who perspire and breathe much, as, for example, workmen or foot-travellers, must take it in greater quantities.

The third reason of its importance lies in this, that it gives us much of the salts and other ingredients that are dissolved in it, and which the human body needs for its support. Those are wrong, therefore, who prefer cistern or distilled water to spring-water; the former being, as it were artificially, free from all metallic and mineral parts which are so beneficial to our health; while spring-water contains them in abundance, and ought, therefore, to be taken in preference even to the purest rain-water.

But one of the most excellent qualities of water is, that one can scarcely ever drink too much of it. If but for a moment in the stomach, it is absorbed there and goes immediately into the blood. From this arises its rapid cooling quality; which, however, may become very dangerous when one is heated. There is but one case in which water is not readily absorbed by the stomach; when it contains salts that make it heavier than blood, for example, Glauber's salt and bitter-salt It passes then into the intestinal canal, and produces here—partly as liquid, partly by its salts exciting the nerves of the intestines—that medicinal effect for which it is famous. Many water-cures, especially those applied in cases of abdominal diseases, are of similar effects.

Common water, however, which is immediately transmitted to the blood, effects by this accelerated secretion of perspiration, respiration and urine ; this constitutes the beneficial effects of water-cures, where a glass of water often produces better results than a bottle of medicine.

If we can control our thirst until several hours after dinner have passed, a glass of beer wlll be a welcome beverage to us. Beer contains nutriment ; it includes more or less albumen, sugar, gluten, hops, and alcohol. Owing to the variety in its fermentation and manufacture, we have many kinds of beer, such as, for example, porter, ale, and, above all others, the lager-beer.

Good beer—that is, beer well brewed and containing all the ingredients this beverage generally does contain—is, very justly, often given to nurses and mothers, because it assimilates easily and very rapidly. It is a kind of soup ; one may take it when a person is too heated or fatigued to eat a regular meal. There is a kind of beer that contains more hops, and is therefore very bitter ; it is very good for the stomach. The Bavarian beer, when genuine, contains more alcohol than the other, which gives it the advantages of liquor without its disadvantages. It therefore does not satisfy one's appetite, but, on the contrary, tends to increase it ; thus it is more adapted to be taken at breakfast and supper. Another kind of beer, called white-beer, contains more sugar and oxygen ; it may, for this reason, supply the place of sugar, and Seltzer-water, and is recommended to all those who need Seidlitz powders.

In another part of this work we shall perhaps speak more about the usefulness of beer. To-day we must pray our readers to be satisfied with what we have said about it ; we shall now speak about supper.

## CHAPTER XIX.

### SUPPER.

No time of the day is more pleasant than the evening hours after the day's work is over; there is a solemn calm and quiet in them which charms both soul and body.

This time of ease and rest must not be disturbed on our part by overburdening the stomach. We eat only for the purpose of compensating for the loss experienced through our work; we should not eat more than is necessary to supply the strength lost; in other words, to give us sufficient strength to continue our labor. And as the day's work is finished, there being not much work before us, we need not take much food.

When glancing at a sleeping person and noticing his long breathing and increased perspiration, one may be led to the belief that he loses much oxygen and water during his sleep; that therefore we must provide ourselves abundantly with food before retiring to bed. This is, however, a mistake. The breath of a sleeping person is long and deep, but very slow; and his perspiring does not cause any great loss of water, but comes rather from this, that one's body during the night is more protected by covers and closed windows, etc., from draft which dries our evaporation, and therefore prevents perspiration in day-time. During sleep we need even less of bodily strength than through the day; for this reason we feel no hunger in the night, and, in spite of the long fasting, no fatigue in the morning.

From this we conclude that supper should not be a meal

for the night, but merely for the last hours of the day. It should be no meal *prænumerando*, but *postnumerando!*

It is therefore best to choose but light dishes, which, if we wish to rest well, must be easily digested, and eaten at least two or three hours before bed-time.

For healthy people a warm supper is unnecessary; our dinner is taken warm for the purpose only of keeping the gluten and fat of the food liquid; as this kind of food, however, is not proper for supper, we need not take it at all in the evening. If we do, it is but an additional burden to the housewife, who surely has enough trouble and labor in the kitchen during the day. He who is not satisfied with a piece of bread and butter and a glass of beer, may eat a piece of cheese besides; but it must be no other kind than sour cheese—the Germans call it *Schmierkaese*—common cheese being too heavy for night because of its containing fat. This sour cheese, whether soft or hardened, is easily digested; it even excites the stomach like spice, especially if you eat it with caraway seeds, and thus promotes the secretion of gastric juice. The other kind of cheese is, for no other reason than that, often eaten after dinner; for, though taken by itself scarcely digestible, if eaten in very small quantity, it increases by its action upon the stomach, the quantity of gastric juice there, and, therefore, promotes digestion in general.

Should we, however, for one reason or the other, insist upon having a more substantial supper, then let us take soft-boiled eggs. The nutritive quality of eggs is equivalent to that of meat. They unite all good sides of the meat; nay, we may say here, that the most nourishing part in meat is nothing but egg-white, or, as we call it, "albumen."

We recommend soft-boiled eggs, because hard ones are difficult to digest. They are best prepared by boiling, if the water is allowed to boil first and the eggs put in afterwards. The reason of this is, that the boiling water hardens the outer part of the egg very rapidly, forming a thick crust, which prevents the heat of the boiling water from penetrating farther.

It is a custom of our country to take tea in the evening. Tea is no article of food, but it possesses the qualities of coffee ; it warms the blood, increases the activity of the heart, and produces a certain freshness of the mind, which is a good remedy against ennui and sleepiness in a company or party.

And since we are speaking of ennui and sleepiness, we think it advisable to close our present subject, "The Articles of Food for the People," and we part from our readers with the full conviction that they will enjoy their real "articles of food" much better than they have relished these scientific conversations about them.